Jennifer Niessner

The role of interfacial area in two-phase flow in porous media

Jennifer Niessner

The role of interfacial area in two-phase flow in porous media

bridging scales and coupling models

Südwestdeutscher Verlag für Hochschulschriften

Impressum/Imprint (nur für Deutschland/only for Germany)
Bibliografische Information der Deutschen Nationalbibliothek: Die Deutsche Nationalbibliothek verzeichnet diese Publikation in der Deutschen Nationalbibliografie; detaillierte bibliografische Daten sind im Internet über http://dnb.d-nb.de abrufbar.
Alle in diesem Buch genannten Marken und Produktnamen unterliegen warenzeichen-, marken- oder patentrechtlichem Schutz bzw. sind Warenzeichen oder eingetragene Warenzeichen der jeweiligen Inhaber. Die Wiedergabe von Marken, Produktnamen, Gebrauchsnamen, Handelsnamen, Warenbezeichnungen u.s.w. in diesem Werk berechtigt auch ohne besondere Kennzeichnung nicht zu der Annahme, dass solche Namen im Sinne der Warenzeichen- und Markenschutzgesetzgebung als frei zu betrachten wären und daher von jedermann benutzt werden dürften.

Coverbild: www.ingimage.com

Verlag: Südwestdeutscher Verlag für Hochschulschriften GmbH & Co. KG
Dudweiler Landstr. 99, 66123 Saarbrücken, Deutschland
Telefon +49 681 37 20 271-1, Telefax +49 681 37 20 271-0
Email: info@svh-verlag.de

Zugl.: Stuttgart, Universität, Habilitation, 2011

Herstellung in Deutschland:
Schaltungsdienst Lange o.H.G., Berlin
Books on Demand GmbH, Norderstedt
Reha GmbH, Saarbrücken
Amazon Distribution GmbH, Leipzig
ISBN: 978-3-8381-2778-1

Imprint (only for USA, GB)
Bibliographic information published by the Deutsche Nationalbibliothek: The Deutsche Nationalbibliothek lists this publication in the Deutsche Nationalbibliografie; detailed bibliographic data are available in the Internet at http://dnb.d-nb.de.
Any brand names and product names mentioned in this book are subject to trademark, brand or patent protection and are trademarks or registered trademarks of their respective holders. The use of brand names, product names, common names, trade names, product descriptions etc. even without a particular marking in this works is in no way to be construed to mean that such names may be regarded as unrestricted in respect of trademark and brand protection legislation and could thus be used by anyone.

Cover image: www.ingimage.com

Publisher: Südwestdeutscher Verlag für Hochschulschriften GmbH & Co. KG
Dudweiler Landstr. 99, 66123 Saarbrücken, Germany
Phone +49 681 37 20 271-1, Fax +49 681 37 20 271-0
Email: info@svh-verlag.de

Printed in the U.S.A.
Printed in the U.K. by (see last page)
ISBN: 978-3-8381-2778-1

Copyright © 2011 by the author and Südwestdeutscher Verlag für Hochschulschriften GmbH & Co. KG and licensors
All rights reserved. Saarbrücken 2011

Contents

1 Motivation — 1
  1.1 General overview of flow and transport in permeable media — 1
  1.2 Multi-scale multi-physics aspects — 4
  1.3 Historical development of models for two-phase flow in porous media — 8
  1.4 Need for a thermodynamically consistent model that includes interfaces — 10
  1.5 Summary — 14
  1.6 Outline of this thesis — 15

2 Classical two-phase flow theory — 16
  2.1 Definition of scales — 18
  2.2 Definition of basic fluid and material properties — 22
    2.2.1 Mass and mole fractions, concentrations — 22
    2.2.2 Density — 23
    2.2.3 Viscosity — 23
    2.2.4 Intrinsic and relative permeability — 24
    2.2.5 Capillary pressure — 26
    2.2.6 Internal energy, enthalpy, and heat capacity — 28
    2.2.7 Thermal conductivity — 29
  2.3 Balance equations — 30
    2.3.1 Mass balance — 31
      2.3.1.1 Two-phase flow — 31
      2.3.1.2 Two-phase–two component flow and transport — 33
    2.3.2 Momentum balance — 34
      2.3.2.1 General momentum balance — 34
      2.3.2.2 Darcy's law and extensions — 36
    2.3.3 Energy balance — 39
  2.4 Constitutive relationships and equations of state — 41
    2.4.1 Density — 41
    2.4.2 Viscosity — 42
    2.4.3 Internal energy and enthalpy — 43
    2.4.4 Equilibrium composition — 43
    2.4.5 Relative permeability — 46
    2.4.6 Capillary pressure — 48
  2.5 Summary — 56

3 Alternative approach—an interfacial-area-based model — 58
- 3.1 Introduction 58
  - 3.1.1 Overview of problematic issues of the classical two-phase flow approach 59
  - 3.1.2 Overview of alternative approaches 63
- 3.2 Balance equations for two-phase flow 65
  - 3.2.1 Basic ideas 66
    - 3.2.1.1 Volume averaging of general phase balances 69
    - 3.2.1.2 Volume averaging of general interface balances . . . 70
  - 3.2.2 Mass balance 71
  - 3.2.3 Momentum balance 72
  - 3.2.4 Energy balance 73
  - 3.2.5 Entropy balance 75
- 3.3 Constitutive relationships 77
  - 3.3.1 Relative permeability 84
  - 3.3.2 Capillary pressure 88
- 3.4 Interphase mass transfer 92
- 3.5 Interphase energy transfer 94
- 3.6 Summary 104

4 Numerical modeling — 106
- 4.1 Overview of space discretization methods and mathematical formulations 107
  - 4.1.1 Discretization methods 107
  - 4.1.2 Fully coupled versus fractional flow formulation 109
    - 4.1.2.1 Fully coupled formulation 110
    - 4.1.2.2 Fractional flow formulation 112
- 4.2 Time discretization 116
- 4.3 Space discretization: vertex centered finite volume method 118
- 4.4 Summary 121

5 Modeling examples — 122
- 5.1 Isothermal immiscible two-phase flow 122
  - 5.1.1 Constitutive relationships 124
    - 5.1.1.1 Specific interfacial area 125
    - 5.1.1.2 Production/destruction rate of specific interfacial area E_{wn} 125
  - 5.1.2 Numerical model 128
  - 5.1.3 Example: primary drainage 129
- 5.2 Capillary redistribution 130
  - 5.2.1 Mathematical approach 132
    - 5.2.1.1 The dimensionless form 134
    - 5.2.1.2 The semi-analytical solution 135
    - 5.2.1.3 An example 143

	5.2.2	Experimental approach	148
	5.2.3	Numerical approach	150
5.3	Kinetic interphase mass transfer		154
	5.3.1	Classical approaches	156
		5.3.1.1 DNAPL pool dissolution	156
		5.3.1.2 Gas–water systems	157
	5.3.2	Alternative approach including interfaces	159
	5.3.3	Numerical test case	162
5.4	Kinetic interphase energy transfer		165
	5.4.1	Background	165
	5.4.2	First numerical example: drying of a porous medium	170
	5.4.3	Second numerical example: evaporator	174
	5.4.4	Discussion	179
5.5	Material interfaces		180
	5.5.1	Continuity conditions	180
	5.5.2	An interface condition	181
5.6	Gas–water processes in a fracture–matrix system		184
	5.6.1	Calculation of effective parameters	186
		5.6.1.1 Porosity	186
		5.6.1.2 Intrinsic permeability	186
		5.6.1.3 Specific interfacial area–capillary pressure–saturation surface	188
	5.6.2	Macro-scale simulation of two-phase flow in a fracture–matrix system	194
5.7	Determination of parameters		200
	5.7.1	Overview of the methods	201
	5.7.2	Examples	202
		5.7.2.1 Flow cell experiments	202
		5.7.2.2 Pore-network models	209
		5.7.2.3 Lattice-Boltzmann simulations	209
5.8	Summary		210

6 Multi-scale–multi-physics modeling: a perspective   212

6.1	Overview of multi-scale multi-physics approaches	213
6.2	Multi-scale multi-physics based on dimensional analysis	216
	6.2.1 Hysteresis	216
	6.2.2 Kinetic interphase mass transfer	217
	6.2.3 Kinetic interphase energy transfer	220
6.3	An example: carbon dioxide storage in the subsurface	224
6.4	Summary	228

7 Summary and conclusions   230

7.1	Summary	230
7.2	Outlook	233

List of Figures

1.1 Different examples for flow and transport processes in geological, biological, and technical porous media. 2
1.2 Carbon dioxide storage in the subsurface: variation of dominant processes in space and time. 5
1.3 Brain cancer treatment by convection-enhanced delivery: multi-scale multi-physics aspects. 6
1.4 Flow and transport through a PEM fuel cell: multi-scale multi-physics aspects. 7
1.5 Schematic representation of spatial and temporal scales. 7
1.6 Idea of multi-scale multi-physics approaches. 8
1.7 Hysteresis of capillary pressure—static case. 12
1.8 Hysteresis of capillary pressure—dynamic case 12
1.9 Interphase mass and energy transfer. 13

2.1 Interfacial tension between water phase and gas phase. 17
2.2 Definition of scales. 19
2.3 Definition of the representative elementary volume. 20
2.4 Wetting angle α between a wetting and a non-wetting fluid. 21
2.5 Fluid–fluid interface in a capillary tube. 27
2.6 Control volume for the phase mass balance. 32
2.7 Control volume for the component mass balance. 33
2.8 Control volume on the pore scale for setting up the pore-scale momentum balance. 35
2.9 Setup of Henry Darcy's original experiment. 37
2.10 Applicability of Henry's law and Raoult's law for a binary gas–liquid system (after Lüdecke and Lüdecke [2000]). 46
2.11 Relative permeability–saturation relationships according to the Brooks & Corey and the van Genuchten model. 48
2.12 Capillary pressure–saturation relationships according to the Brooks & Corey and the van Genuchten model. 51
2.13 Capillary hysteresis for a porous medium that is initially fully wetting-phase saturated. 52
2.14 Impact of dynamic effects on $p_n - p_w$–saturation relationships. 55

3.1 Hysteresis of capillary pressure: all (S_w, p_c) values between main imbibition and main drainage curve are physically possible. 62

3.2	In the dynamic case, (S_w, p_c) values even outside the main hysteresis loop are possible.	62
3.3	Interphase mass and energy transfer.	63
3.4	Definition of phases, interfaces, and common lines.	67
3.5	Wetting fluid film on the solid surface.	67
3.6	Two fluids in a capillary tube—molecular-scale picture.	67
3.7	Exemplary $a_{wn}(S_w, p_c)$ surface with scanning curves in the $p_c(S_w)$ plane (taken from Nuske [2009]).	91
3.8	Conceptual models with respect to heat and energy transfer.	95
3.9	Lattice-Boltzmann simulation of a drainage process in a porous medium.	101
3.10	Relationship $a_{ws}(S_w, p_c)$ as obtained by Ahrenholz et al. [2009] from Lattice-Boltzmann simulations of air–water flow in a natural porous medium.	102
3.11	Relationship $a_{ns}(S_w, p_c)$ as obtained by Ahrenholz et al. [2009] from Lattice-Boltzmann simulations of air–water flow in a natural porous medium.	103
3.12	Estimation of a_{ws} and a_{ns} at a given p_c from an $a_{wn}(S_w, p_c)$ function.	104
4.1	Construction of control volumes for a finite element grid.	118
5.1	Relation between specific interfacial area, saturation, and capillary pressure.	126
5.2	Boundary and initial conditions for the drainage problem (example 1).	129
5.3	Distribution of S_w and p_c in example 1 using the 2p model and of S_w, p_c, and a_{wn} using the 2pia model.	131
5.4	Setup of the redistribution problem.	132
5.5	The capillary pressures: the outer left and right curves are the bounding imbibition and drainage curves.	146
5.6	The imbibition (left) and drainage (right) fluxes as functions of the saturation. Here $S^+ = 0.365$ and $S^- = 0.498$.	147
5.7	The selfsimilar saturation in the imbibition (left) and drainage (right) subdomains.	147
5.8	The selfsimilar capillary pressures in the imbibition (left) and drainage (right) subdomains.	148
5.9	Saturation distribution and capillary pressure–saturation curves in an infinite porous medium.	149
5.10	Saturation distribution and capillary pressure–saturation curves in a semi-infinite or finite porous medium.	149
5.11	Gamma-system with robot for horizontal movement and measuring column.	151
5.12	Profiles of dimensionless saturation $S(\eta)$ over time.	152
5.13	Profiles of dimensionless capillary pressure $p(\eta)$ over time.	153
5.14	Profiles of dimensionless specific interfacial area $a(\eta)$ over time.	153

5.15 Setup of the test example. 163
5.16 Contour plots of S_w, p_n, $\bar{X}_w^w$, $\bar{X}_n^a$, p_c, a_{wn} at four selected time steps. . . 165
5.17 Conceptual models with respect to heat and energy transfer on the pore and macro scale using a classical approach and using an interfacial-area-based model. 166
5.18 Setup of the numerical example. 172
5.19 Contour lines of a_{wn} for the new approach and S_w for new and classical approach after 2 and 4 s. 172
5.20 Contour lines of X_n^w and X_w^a for new and classical approach after 2 and 4 s. 173
5.21 Contour lines of T for the classical approach and T_n as well as T_w-T_n, $T_n - T_s$ for the new approach after 2 and 4 s. 173
5.22 Setup of the numerical example: water passes a heat source and is evaporated. 174
5.23 Water saturation and water–gas specific interfacial area after 17 s. . . 175
5.24 Mass fractions of air in water after 17 s. 176
5.25 Mass fractions of vapor in the gas phase after 17 s. 177
5.26 Temperatures of the phases after 17 s using the interfacial-area-based model and the classical model. 178
5.27 Temperature differences between wetting and non-wetting phase and between non-wetting and solid phase. 178
5.28 Vertex centered finite volume discretization using the 2pia model (1d consideration). 181
5.29 Graphs of $p_c(S_w)$ for two points located directly at a material interface and $a_{wn}(S_w)$ for a fixed capillary pressure for fine and coarse material. 182
5.30 From the conceptual model of a single fracture over micro-scale simulation to macro-scale simulation. 186
5.31 Different representations of the fracture surfaces. 187
5.32 Simplified interface geometry. 189
5.33 Illustration of a drainage process. 191
5.34 Water and gas in three fracture elements. 192
5.35 Schematic of the connectivity situation for an imbibition process. . . 193
5.36 Sketches of the applied model setups. 194
5.37 Comparison of the different model concepts for the homogeneous case. 196
5.38 Capillary pressure and mass fraction X_w^a as a function of time at a node. 197
5.39 Mass transfer after 50 seconds. 197
5.40 Mass transfer depending on infiltration rates 198
5.41 Grid dependence of the solution. A simulated time of 6 seconds is compared. 199
5.42 Comparison of mass transfer for a vertical and an inclined fracture. . 200
5.43 Flow cell and cross-sections of flow cell matrix. 203
5.44 Notation used for the calculation of interfacial permeability K_{wn}. . . 206

5.45 a) Wetting-phase saturation, b) capillary pressure, c) specific interfacial area as a function of time, and d) specific interfacial area as a function of wetting-phase saturation for $n = 3$. 208
5.46 Example of a three-dimensional pore network consisting of cylindrical pore throats and spherical pore bodies. 209
5.47 Identification of interfacial areas from results of a three-dimensional multi-phase Lattice-Boltzmann simulation. 210

6.1 Schematic representation of spatial and temporal scales. 212
6.2 Solubility limits $X_{s,w}^a$ and $X_{s,n}^w$ and actual mass fractions $\bar{X}_w^a$ and $\bar{X}_n^w$ along the line $y = 0.25$ m for two different time steps and five different Damköhler numbers. 219
6.3 Carbon dioxide injection into the subsurface. 224
6.4 Hysteresis due to end of injection. 225
6.5 Hysteresis due to heterogeneities. 225
6.6 Hysteresis due to heterogeneities: pooling and trapping. 226
6.7 Chemical non-equilibrium. 227
6.8 Thermal non-equilibrium. 228
6.9 Schematic of the multi-scale algorithm for the example of CO_2 storage. 229

List of Tables

5.1 Parameters for the drainage problem of example 1. 130
5.2 "Classical" parameter values for the numerical example. 171
5.3 Additional parameters for the numerical example using the interfacial-area-based model approach. 171
5.4 Coefficients of the specific interfacial area–capillary pressure–saturation surface for matrix (Joekar-Niasar et al. [2008]) and fracture (Nuske [2009]). 195
5.5 Initial values and Dirichlet boundary conditions 196
5.6 Computational expenses for a 100 second simulation run. 198

Nomenclature

Latin symbols

Symbol	Meaning	Unit
$a_{\alpha\beta}$	specific interfacial area of $\alpha\beta$-interface	$\frac{1}{m}$
a_{ij}	coefficients in interfacial area function	various
b_i	coefficients	various
$d_\alpha^T, d_{\alpha\beta}^T$	thermal diffusion length of phase α or $\alpha\beta$-interface	m
c	specific heat capacity	$\frac{J}{kg \cdot K}$
d	diameter	m
d^κ	diffusion length of component κ	m
$\hat{e}$	mass transfer function	$\frac{1}{s}$
e_{wn}	production function of specific interfacial area	$\frac{1}{m}$
f_α	fractional flow function of phase α	—
g	gravity	$\frac{m}{s^2}$
h	piezometric head	m
h_α	specific enthalpy of phase α	$\frac{J}{kg}$
j_α^κ	macro-scale diffusive flux of component κ in phase α	$\frac{kg \cdot m^2}{s}$
k	mass transfer rate coefficient	$\frac{m}{s}$
k_f	hydraulic conductivity	$\frac{m}{s}$
$k_{r\alpha}$	relative permeability of phase α	—
m	mass, van Genuchten parameter	kg, —
n	number of moles	mole
$\underline{n}_{\alpha\beta}$	vector normal to $\alpha\beta$-interface pointing out of α into β	—
p_α	pressure of phase α	Pa
p_c	capillary pressure	Pa
p_{sat}^w	saturation vapor pressure for water	Pa
p_d	entry pressure	Pa
$\underline{q}$	partial macroscopic heat vector	$\frac{kg}{s^3}$
t	time	s
$\underline{\underline{t}}$	stress tensor	$\frac{kg}{m \cdot s^2}$
u	velocity component, specific internal energy	$\frac{m}{s}, \frac{J}{kg}$
v	specific volume	$\frac{m^3}{kg}$
$\underline{v}_\alpha, \underline{v}_{\alpha\beta}$	velocity of phase α or $\alpha\beta$-interface	$\frac{m}{s}$
x_α^κ	mole fraction of component κ in phase α	—
$\underline{x}$	spatial vector	m
A	flux term, area, Helmholtz free energy density	$\frac{kg}{m^3 \cdot s}, m^2, \frac{m^2}{s^2}$
C	concentration	$\frac{kg}{m^3}$
$C_{p,\alpha}, C_{p,\alpha\beta}$	specific heat capacity of phase α or $\alpha\beta$-interface, respectively	$\frac{J}{kg \cdot K}$
C_{ia}	interfacial number	—
$\underline{\underline{D}}_\alpha^{th}$	thermal dispersion tensor of phase α	$\frac{m^2}{s}$

Symbol	Description	Units
Da^κ	Damköhler number of component κ	—
D^κ	micro-scale diffusion coefficient of component κ	$\frac{m^2}{s}$
D_α^κ	macro-scale diffusion coefficient of component κ in phase α	$\frac{m^2}{s}$
E_{wn}	production rate of specific interfacial area	$\frac{1}{ms}$
F	flux	various
F_α	external sink / source of phase α	$\frac{m^3}{s}$
$G_\alpha, G_{\alpha\beta}$	specific Gibbs free energy of phase α or $\alpha\beta$-interface	$\frac{m^2}{s^2}$
$\underline{G}$	gravity term	$\frac{Pa}{m}$
H	enthalpy	J
H_{w-n}^κ	Henry coefficient of component κ with respect to w and n phase	$\frac{1}{Pa}$
I	mass exchange term	$\frac{kg}{m^3 \cdot s}$
$\underline{\underline{K}}$	intrinsic permeability tensor	m^2
$\underline{\underline{K}}_{wn}$	interfacial permeability tensor	$\frac{m^3}{s}$
K_M	mean curvature of an interface	$\frac{1}{m}$
L	characteristic length	m
L_α^{th}	characteristic thermal diffusion length of phase α	m
M	molar mass, storage term	$\frac{kg}{mole}, \frac{kg}{m^3 \cdot s}$
M_α	mass ratio with respect to phase α	—
N	basis function	—
$Nu_{\alpha\beta}$	Nusselt number for heat transfer from phase α to phase β	—
Pe_α	Peclet number of phase α	—
Q	water flow rate	$\frac{m^3}{s}$
Q_α^κ	external source or sink of component κ in phase α	$\frac{1}{s}$
R	radius of curvature, individual gas constant	m, $\frac{J}{kg \cdot K}$
$\underline{\underline{R}}$	resistance tensor	$\frac{1}{m^2}$
Re	Reynolds number	—
S_α	saturation of phase α	—
Sh	Sherwood number	—
T	temperature	K
$\hat{\underline{T}}$	momentum transfer	$\frac{N}{m^3}$
U	internal energy	J
V	volume	m^3
W	work, weighting function	J, —
X_α^κ	mass fraction of component κ in phase α	—
$X_{\alpha,s}^\kappa$	equilibrium mass fraction of component κ in phase α	—

Greek and other symbols

Symbol	Meaning	Unit
α	dispersivity, wetting angle, van Genuchten parameter	m, °, $\frac{1}{Pa}$
δ_{ij}	Kronecker symbol	–
ε	tolerance	–
η	specific entropy, similarity variable	$\frac{m^2}{s^2 K}$, $\frac{m}{s^{1/s}}$
γ	macro-scale interfacial tension	$\frac{N}{m}$
λ	Brooks-Corey parameter, mobility, thermal conductivity	–, $\frac{1}{Pa \cdot s}$, $\frac{W}{m \cdot K}$
μ_α	dynamic viscosity of phase α	$Pa \cdot s$
ν_α	kinematic viscosity of phase α	$\frac{m^2}{s}$
ϕ	porosity	–
ψ	general physical property, total potential	various, Pa
ρ_α	density of phase α	$\frac{kg}{m^3}$
σ	micro-scale interfacial tension	$\frac{N}{m}$
$\mathcal{I}_\alpha, \mathcal{I}_{\alpha\beta}$	friction forces acting on phase α, or $\alpha\beta$-interface	$\frac{kg}{m^2 s^2}$
τ	tortuosity, dynamic parameter	–, $Pa \cdot s$
$\Gamma_{\alpha\beta}$	areal mass density of $\alpha\beta$-interface	$\frac{kg}{m^2}$
Λ	entropy production	$\frac{J}{K \cdot s}$
$\underline{\underline{T}}$	deviatoric stress tensor	$\frac{kg}{m \cdot s^2}$
$\mathcal{R}$	universal gas constant	$\frac{J}{mole \cdot K}$
$< \cdot >$	averaged quantity	–
$'$	deviation quantity	–

Subscripts

Symbol	Meaning
c	capillary
e	effective
n	non-wetting
p	at constant pressure
pm	porous medium
s	solid, solubility limit
tot	total
v	at constant volume
w	wetting
wn, tot	total wn-interfaces
L	longitudinal
R	reference
T	transversal, at constant temperature
α	phase

$\alpha\beta$	interface
α, s	equilibrium value in phase
α, Q	external source property of phase α

Superscripts

Symbol	Meaning
a	air
sat	saturation
w	water
th	thermal
FU	fully upwind
T	thermal
κ	component
*	(indicates dimensionless quantity)

1 Motivation

1.1 General overview of flow and transport in permeable media

Fluid flow and transport in permeable media is a subject of general relevance in science, technology, and environment. Systems under consideration include geological, biological, and technical systems. Classical applications comprise petroleum engineering where oil and gas are produced from porous sedimentary rock, civil engineering where stability issues of geotechnical systems are investigated, and environmental engineering where flow and transport of contaminants in the subsurface is considered. Recently, new disciplines dealing with flow and transport in porous and fractured media have evolved. They include the study of the storage of carbon dioxide in deep geological formations, the investigation of the migration of methane emitted by abandoned coal mines, paper manufacturing, study of processes in polymer electrolyte membrane fuel cells, and medical applications, such as the description of fluid flow processes in the human body. Selected examples are shown in Fig. 1.1 and subdivided into examples for geological, biological, and technical systems. In the following, these examples will be briefly discussed.

a) geological systems

- **groundwater flow.** The quantitative prediction of groundwater flow is essential for the extraction of drinking water. The picture shows the drawdown cone of a groundwater extraction well.

- **carbon dioxide storage.** In order to mitigate the greenhouse effect, carbon dioxide is injected into deep geological formations, e.g. below a dome-shaped caprock. In order to determine the storage ability, advective–diffusive dis-

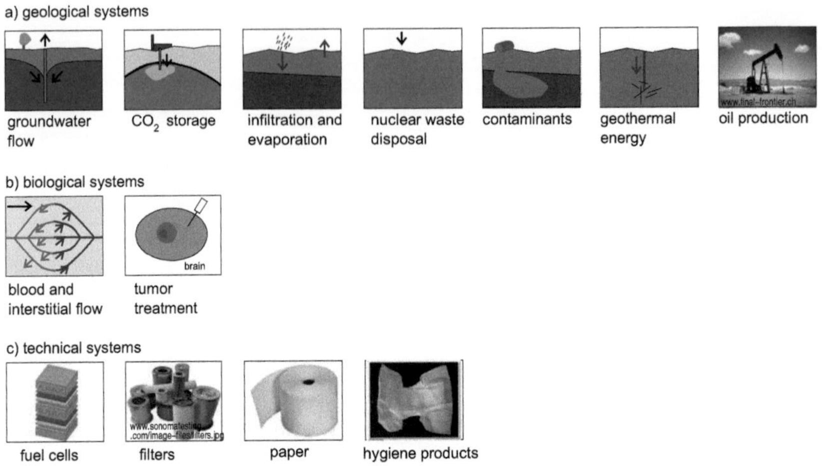

Figure 1.1: Different examples for flow and transport processes in geological, biological, and technical porous media.

placement, dissolution, and mineralization of carbon dioxide are relevant processes.

- **infiltration and evaporation processes.** The upper layer of the subsurface represents the coupling element to the atmosphere. Both infiltration of rain and evaporation processes determine the water balance of that layer and are thus crucial for agriculture and other essentially important forms of land use.

- **nuclear waste disposal.** For the disposal of atomic waste, often, an "impermeable" storage site deep below the surface is chosen, e.g. a salt cavern. However, it may happen that the storage bins corrode and radioactive gases may evade. This effect is a crucial flow and transport process as the highly hazardous gases may quickly migrate to the land surface.

- **contaminant movement and remediation.** Contaminations from various sources and of different nature may escape into the subsurface. In order to effectively apply remediation techniques, it is crucial to know where the contaminant is transported in the subsurface.

- **extraction of geothermal energy.** Heat stored in the earth can be used directly for heating purposes or it can be utilized to generate electrical power. Different

1 Motivation

1.1 General overview of flow and transport in permeable media

Fluid flow and transport in permeable media is a subject of general relevance in science, technology, and environment. Systems under consideration include geological, biological, and technical systems. Classical applications comprise petroleum engineering where oil and gas are produced from porous sedimentary rock, civil engineering where stability issues of geotechnical systems are investigated, and environmental engineering where flow and transport of contaminants in the subsurface is considered. Recently, new disciplines dealing with flow and transport in porous and fractured media have evolved. They include the study of the storage of carbon dioxide in deep geological formations, the investigation of the migration of methane emitted by abandoned coal mines, paper manufacturing, study of processes in polymer electrolyte membrane fuel cells, and medical applications, such as the description of fluid flow processes in the human body. Selected examples are shown in Fig. 1.1 and subdivided into examples for geological, biological, and technical systems. In the following, these examples will be briefly discussed.

a) geological systems

- **groundwater flow.** The quantitative prediction of groundwater flow is essential for the extraction of drinking water. The picture shows the drawdown cone of a groundwater extraction well.

- **carbon dioxide storage.** In order to mitigate the greenhouse effect, carbon dioxide is injected into deep geological formations, e.g. below a dome-shaped caprock. In order to determine the storage ability, advective–diffusive dis-

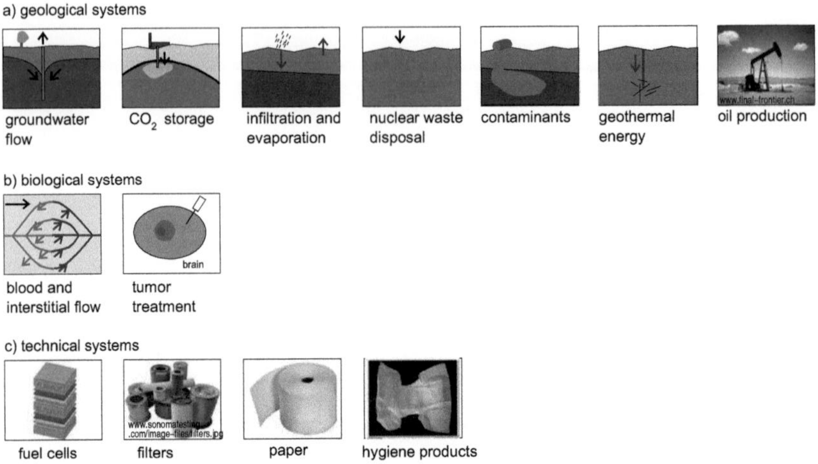

Figure 1.1: Different examples for flow and transport processes in geological, biological, and technical porous media.

placement, dissolution, and mineralization of carbon dioxide are relevant processes.

- **infiltration and evaporation processes.** The upper layer of the subsurface represents the coupling element to the atmosphere. Both infiltration of rain and evaporation processes determine the water balance of that layer and are thus crucial for agriculture and other essentially important forms of land use.

- **nuclear waste disposal.** For the disposal of atomic waste, often, an "impermeable" storage site deep below the surface is chosen, e.g. a salt cavern. However, it may happen that the storage bins corrode and radioactive gases may evade. This effect is a crucial flow and transport process as the highly hazardous gases may quickly migrate to the land surface.

- **contaminant movement and remediation.** Contaminations from various sources and of different nature may escape into the subsurface. In order to effectively apply remediation techniques, it is crucial to know where the contaminant is transported in the subsurface.

- **extraction of geothermal energy.** Heat stored in the earth can be used directly for heating purposes or it can be utilized to generate electrical power. Different principles are available. In the case shown here, cool water is injected into the subsurface which creates an artificial fracture system. From a neighboring well, hot water is extracted.

- **oil production.** Oil is a fossil fuel and serves for the production of electricity and as a fuel for almost all means of transport. Additionally, oil is used in the chemical industry for the production of plastic and other chemical products. The processes occuring during oil production involve the flow of at least two fluid phases (oil and water) in a porous medium. For secondary and tertiary oil recovery, even a third phase (gas) is involved.

b) **biological systems**

- **blood and interstitial flow.** A highly important effect in creatures is the transfer of substances from the blood vessels to the interstitial space and vice versa. This is crucial for delivery of therapeutic agents to the relevant places in the body. The blood vessels represent a "fracture" system with blood as a non-Newtonian fluid and the interstitial space can be regarded as a deformable porous medium filled by the Newtonian interstitial liquid. The picture shows the capillary blood system in an organ. Arterioles supply the organs with oxygen and nutrients while venoles carry away waste products and blood with a lower oxygen content.

- **tumor treatment.** Tumors are treated in different ways depending on where the tumor is located. For brain tumors, for example, the therapeutic agent is injected directly into the interstitial space. This procedure is called convection-enhanced delivery.

c) **technical systems**

- **fuel cells.** Polymer Electrolyte Membrane Fuel Cells (PEM fuel cells) are widely used as mobile sources of energy. Flow and transport processes within the full cell need to be known in detail for the effective construction and optimization of the cells.

- **filters.** Filters are often constructed as porous membranes or layers that are designed to retain certain objects or substances while letting others pass. Filters are often used to remove harmful substances from air or water, for example to remove air pollution or purify potable water, but also to prepare beverages, such as coffee and tea.

- **paper.** Paper also represents a porous medium. Both the pulp dewatering during the manufacturing process and the infiltration of e.g. printer ink into paper represent flow and transport processes in porous media.
- **hygiene products.** Hygiene products, such as diapers, are extensively used world-wide. Hence, any improvement of the understanding of the flow and transport processes through these permeable structures can greatly enhance the design and comfort of these devices.

In all of the systems discussed, processes take place on a large variety of spatial and temporal scales. Spatial scales range from a molecular level to the system size which may take dimensions of several kilometers. Temporal scales may be similarly diverse: while reactions or phase transitions occur in the order of tenths or hundredths of seconds, large-scale flow and transport processes may take decades or centuries. This aspect represents the **multi-scale** character of flow and transport in permeable media. But even on one spatial and temporal scale, processes may vary in space and time. A spatial variation of processes occurs if in some parts of a domain of interest, other physical processes occur than in the remaining parts of the domain. A temporal variation of processes, contrarily, is given if in a subdomain of the domain of interest, different processes dominate in different time regimes. This is the **multi-physics** aspect of flow and transport in porous media. Both multi-scale and multi-physics character will be addressed in more detail in the next section.

1.2 Multi-scale multi-physics aspects

Multi-scale multi-physics aspects of flow and transport processes in permeable media are illustrated using three examples, one from each of the compartments (a geological, a biological, and a technical system): 1) carbon dioxide storage in deep geological formations, 2) brain cancer treatment through convection-enhanced drug delivery, and 3) gas–water processes in the cathode diffusion layer of a PEM fuel cell.

1. Carbon dioxide storage.

    Fig. 1.2 shows the temporal development of a sector of the subsurface containing a carbon dioxide storage site, see e.g. IPCC [2005]. During the injection period, the flow is driven by the flow field imposed by the injection well: the

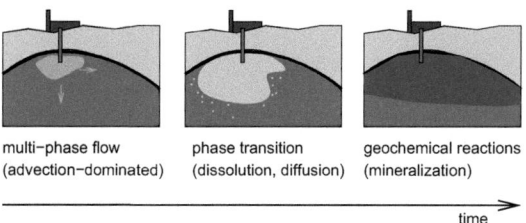

Figure 1.2: Carbon dioxide storage in the subsurface: variation of dominant processes in space and time.

flow is advection-dominated. As CO_2-rich water is heavier than pure water, it will sink down due to gravity forces. At an intermediate time level, the diffusion and dissolution of CO_2 into the water dominate the system. At an even later stage, the CO_2 will be mineralized due to geochemical reactions.

It can be clearly seen that this system is of the multi-physics type both in time and space: Different kinds of processes take place in different time periods after the injection. If e.g. the behavior of the system during or shortly after the injection is of interest it is not necessary to consider geochemical reactions; if, contrarily, the interest is in the long-time period, these geochemical reactions become the dominating processes. The multi-physics character in space becomes obvious by considering Fig. 1.2: processes are most complex exactly in the domain where the CO_2 plume is present. In this domain, two-phase flow occurs at the early time stage, dissolution and diffusion in the intermediate time period, and mineralization in the late time period. Outside of this domain, processes are physically simpler (e.g., single-phase conditions prevail).

The multi-scale character of this system becomes obvious when considering the processes involved, both inside and outside of the CO_2 plume: the processes inside the plume (e.g. dissolution, geochemical reactions) are physically complex and take place on a very small spatial scale. Thus, they need to be resolved on this fine scale. The processes outside the plume are physically less complex (single-phase flow) and may thus be upscaled to a larger scale without significant loss of accuracy. Clearly, this system additionally shows a multi-scale character in time. On a small temporal scale, (almost) immiscible multi-phase flow dominates, at an intermediate time scale, phase transition phenomena play the major role, and finally, on a long temporal scale, geo-

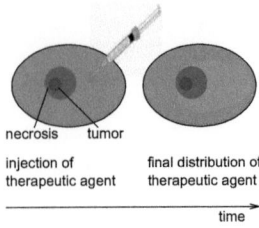

Figure 1.3: Brain cancer treatment by convection-enhanced delivery: multi-scale multi-physics aspects.

chemical reactions represent the dominating processes.

2. Brain cancer treatment.

   Mortality of brain cancer patients is extremely high. This is partially due to the fact that brain tumors cannot be treated with "classical" treatment methods, such as chemotherapy, since the so-called blood–brain barrier prevents all but tiny molecules to leave the blood vessels inside the brain and be transported through the interstitial space towards the tumor. Instead, brain tumors (which are typically surrounded by dead tumor tissue, the necrosis) have to be treated by injection of the therapeutic agent directly into the brain interstitial space as shown on the left hand side of Fig. 1.3 (Smith and Humphrey [2006]). In consequence of the pressure created by the injection, the therapeutic agent is advectively transported towards the tumor.

   As the brain interstitial space is not rigid, the injection will cause a deformation of the porous matrix. This deformation will only be significant in the injection zone. Thus, this system represents a multi-physics system: flow and transport of the therapeutic agents need to be described within the whole brain, while deformation only has to be taken into account locally. Additionally, the distribution of the therapeutic agent outside the injection region is simpler and hence, might be upscaled while the processes directly inside the injection region need very careful consideration and fine resolution.

3. PEM fuel cell.

   PEM fuel cells represent mobile devices for the generation of electrical energy from a controlled reaction of hydrogen and oxygen. In order to highlight the occuring multi-scale multi-physics aspects, we focus our attention on two lay-

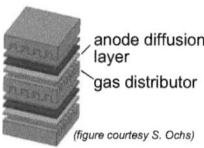

Figure 1.4: Flow and transport through a PEM fuel cell: multi-scale multi-physics aspects.

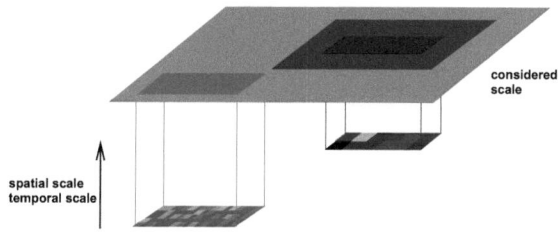

Figure 1.5: Schematic representation of spatial and temporal scales.

ers: the gas distributor and the anode diffusion layer, see Fig. 1.4. In the gas distributor, free flow of gas takes place while in the diffusion layer, porous media flow occurs (Ochs et al. [2007]). This represents the multi-physics character of this system. Additionally, the diffusion layer is very thin compared to the dimensions of the gas distributor. This difference in scale makes a multi-scale method attractive.

In order to get a profound process understanding and to make predictions, for all of the systems above, models for flow and transport processes in porous media are needed. We can generalize the systems discussed as shown in Fig. 1.5 where we see that on the considered scale, different processes occur in different parts of the domain of interest (multi-physics). Additionally, some processes take place on smaller spatial or temporal scales. If an accurate description of the flow and transport processes is desired, the common approach is to first identify the most complex processes occurring in space and time, then to identify the relevant spatial and temporal scales, and finally, to solve the associated numerical problem taking into account the most complex of the considered processes on the finest relevant scale in space and time. In Fig. 1.5, this would correspond to the scale at the bottom.

Recently, new approaches were developed which are tailored to solve models on relevant space and time scales taking relevant processes into account. Using them, different processes on different scales can be considered in different time frames, see

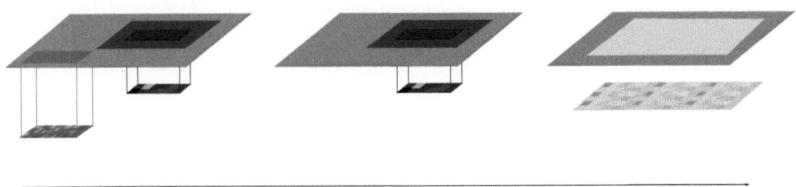

Figure 1.6: Idea of multi-scale multi-physics approaches: consider only relevant processes on relevant scales in relevant domains and time periods.

the exemplary sketch in Fig. 1.6. Here, the vertical axis represents the spatial scales, while the time axis is plotted in the horizontal direction. Obviously, in different time frames, different processes are considered on different scales.

Approaches which allow to account for different processes on different spatial and temporal scales and in different subdomains are commonly called multi-scale multi-physics approaches. Their key ideas are the following

- only relevant processes are considered on each spatial and temporal scale,
- depending on the processes, relevant spatial scales and spatial domains are chosen,
- processes are solved on relevant temporal scales and within relevant time frames.

This procedure allows to tremendously reduce the amount of required data and computing time. A key challenge is, however, how to define error estimators or indicators that allow to choose the appropriate processes, spatial domains and time frames, as well as spatial and temporal scale.

1.3 Historical development of models for two-phase flow in porous media

Many of the systems discussed above include more than one fluid phase (for a precise definition of this term, see the introduction of Chap. 2). A physically based description of these systems is an especially challenging issue.

Such models for two-phase flow in porous media have been developed over the last decades and centuries starting basically with the work of Henry Darcy (Darcy [1856]). He postulated that the flow rate of water through a homogeneous and isotropic porous medium in a quasi one-dimensional column is proportional to the hydraulic head gradient across that column. Buckingham [1907] developed the first multi-phase concept. He introduced unsaturated hydraulic conductivitites that depend on the water content. Almost a quarter century later, Richards [1931] extended the work of Buckingham [1907] by formulating a partial differential equation for water flow in the unsaturated zone of the subsurface which was thereafter called Richards equation. A decade later, Leverett [1941] pioneered in the field of fundamentals of capillarity in porous soils. After that, the interest in two-phase flow in porous media increased tremendously and a number of basic text books were written: one of the most-famous textbooks was written by Bear [1972] and gives a comprehensive introduction into the fundamental fluid dynamics processes and their mathematical description. Another famous book was published around the same time: Scheidegger [1974] addresses the physics of flow and transport through porous media. In parallel, the petroleum industry started to show profound interest in the physics of flow and transport in porous media with the aim of optimizing the exploitation of oil and gas reservoirs. Aziz and Settari [1979], Peaceman [1977] and Chavent and Jaffré [1986] focus on methods to model the flow processes in petroleum reservoirs with numerical simulators, and Lake [1989] discusses in detail the techniques for *Enhanced Oil Recovery* (EOR). Another more general textbook on the modeling of transport phenomena in porous media is the book of Bear and Bachmat [1990]. The book of Looney and Falta [2000] deals in great detail with the flow and transport processes in the vadose zone having also a strong focus on forward and inverse modelling techniques.

The model capabilities improved significantly in the last decades and many models include complex coupled and non-linear multi-phase processes including mass transfer between the fluid phases. This induced a strong need for an accurate prediction of fluid properties, equations of state, and constitutive relationships. Due to the complexity of these systems of equations, there is a strong demand for sophisticated algorithms and discretization methods in order to solve the arising systems of nonlinear partial differential equations in a fast and efficient way.

1.4 Need for a thermodynamically consistent model that includes interfaces

As shown in this short abstract on the historical development of two-phase flow modeling, simple models have been extended to complex models over the decades and centuries. Sometimes, however, these extensions did not have a sound physical base (Rose [2000]) leading to a number of problematic issues and non-physical behavior of the classical two-phase flow equation system. Instead of directly deriving the equations for complex physical systems, existing equations for simple systems were generalized by adding factors and making parameters and constitutive relationships phase- or component-dependent. In order to illustrate this, problematic issues of the classical two-phase flow model are briefly addressed concerning Darcy's Law, macro-scale capillary pressure, and interphase mass and energy transfer. These issues will be discussed in further detail in Sec. 3.1.

1. Darcy's Law
The original Darcy's Law was proposed as an empirical relationship linearly relating the specific flux through a column to the water head difference at the two sides of the column assuming a homogeneous and isotropic porous medium and one-dimensional isothermal flow, see Darcy [1856]. The same form of Darcy's Law is used today to calculate flow velocities of fluid phases in physically highly complex systems, such as for non-isothermal compositional, multi-dimensional multi-phase flow processes in heterogeneous and anisotropic porous media. The only changes are that parameters are made phase and / or component dependent and factors are added. This means that in this physically highly complex situation, still, the only driving forces for flux of a phase are pressure difference and gravity. It has been shown experimentally, that there are indeed further driving forces apart from pressure gradient and gravity. It seems that the effect of other relevant driving forces has to be absorbed by an empirical function commonly called relative permeability such that the linearity between flux and pressure gradient does actually hold. The relative permeability is introduced as a scaling parameter, reducing the permeability of one phase due to the presence of another phase. However, this parameter does not come from any balance equations or an averaging process; instead, it is an empirical function. It is well known that this function is hysteretic. But even worse: it has been shown, both experimentally and numerically, that relative permeability is a function of a list of other parameters, including flow velocity, viscosity ratios,

and boundary conditions (Miller et al. [1998], Demond and Roberts [1987], Avraam and Payatakes [1995a,b], Lefebvre du Prey [1973], Jerauld and Salter [1990])!

2. Macro-scale capillary pressure

Another critical point of classical two-phase flow models is the treatment of capillary pressure. While pore-scale capillary pressure is defined as the pressure drop across fluid–fluid interfaces at equilibrium and is directly related to physical quantities, such as the curvature of the interface and interfacial tension, the meaning and the relation of macro-scale capillary pressure to other variables is not clear a priori. Usually in classical models, macro-scale capillary pressure is defined as the difference between averaged phase pressures. The averaging region or volume corresponds to a representative elementary volume (REV). Thus, this quantity which is from the physical point of view an interface-related parameter, is now defined with respect to a volume. The classical model of two-phase flow consisting of mass balance equations and extended forms of Darcy's laws is not closed a priori; therefore, a closing condition needs to be found. As capillary pressure and saturation are known to be correlated, the classical model is closed by postulating that macro-scale capillary pressure is a function of wetting-phase saturation only. However, this is problematic, since this relationship is highly hysteretic: in numerous experimental studies measuring non-wetting and wetting phase pressure along with saturation under no-flow conditions (static experiments), it has been shown that capillary pressure is dependent on the process (i.e. whether wetting-phase saturation is increasing or decreasing) and on the saturation history. This typically leads to a situation shown in Fig. 1.7: during different drainage (decreasing wetting-phase saturation) and imbibition (increasing wetting-phase saturation) cycles, all capillary pressure–saturation values between a bounding drainage and a bounding imbibition function are possible! Fig. 1.7 brings up the following question: is saturation really the only parameter on which capillary pressure depends?

The situation becomes even worse if one accounts for the fact that most often, dynamic situations are of interest where non-zero flow velocities occur. In that case, experimental studies show that values of capillary pressure (being classically defined as the difference in average phase pressures) *outside* of the bounding drainage and imbibition curves determined under static conditions occur! This means that in principle, all capillary pressure–saturation values in the positive capillary pressure–saturation plane represent physically possible states, see the gray area in Fig. 1.8! This clearly indicates that capillary pressure is a function of not only saturation and

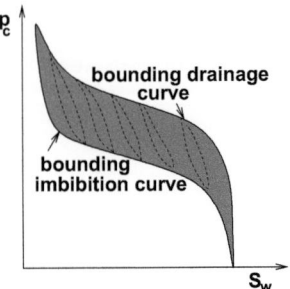

Figure 1.7: Hysteresis of capillary pressure: for static conditions, all wetting phase saturation–capillary pressure values between a bounding imbibition and a bounding drainage curve are physically possible.

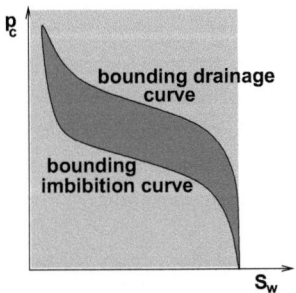

Figure 1.8: In the dynamic case, wetting phase saturation–capillary pressure values even outside the static bounding imbibition and bounding drainage curve are physically possible.

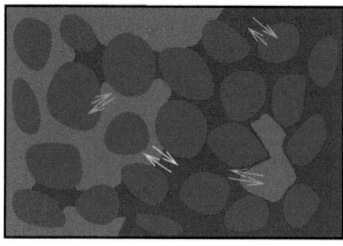
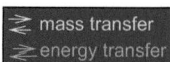

Figure 1.9: Interphase mass and energy transfer.

that the classical approach is—at best—incomplete!

3. Interphase mass and energy transfer

Interphase mass transfer takes place across the fluid–fluid interface. Therefore, if chemical equilibrium is not reached, it is highly dependent on the fluid–fluid interfacial area. Although this fact is widely acknowledged, classical models do not and cannot account for this fact since interfacial area is unknown. That means that it is not possible to account for interphase mass transfer under non-equilibrium conditions in a physically-based manner. Instead, classical models either describe interphase mass and energy transfer by equilibrium relations or they use empirical models to describe the kinetics. Similar considerations can be made with respect to interphase energy transfer. But here, in addition to fluid–fluid interfaces, solid–fluid interfaces need to be taken into account as, unlike mass, energy may also be transferred from and to the solid phase, see Fig. 1.9.

The above questions have led to research activities questioning the physical correctness of classical two-phase flow models. Gray and Hassanizadeh [1991] could even show that the oversimplified concepts and theories of the classical approach lead to a number of paradoxes and inconsistencies. Alternative conceptual and mathematical models have been developed which describe two-phase flow based on thermodynamic principles. Among these are i) mixture theory which has been applied to two-phase flow by Bowen [1982], ii) a rational thermodynamics approach developed by Hassanizadeh and Gray [1980, 1990, 1993a,b], iii) a thermodynamically constrained averaging theory (Gray and Miller [2005], Miller and Gray [2005]), and iv) an approach based on averaging and non-equilibrium thermodynamics by Marle [1981] and Kalaydjian [1987]. Furthermore, Hilfer [2006] suggested a model with separate balance equations for the percolating and non-percolating fraction of each phase. All of these approaches try to describe two-phase flow using a basis more physical than the classical model, and they all try to resolve the three issues discussed above (or parts of them). E.g. the thermodynamically-based model of Hassanizadeh and Gray [1990], unlike the classical approach, not only includes balance equations for the bulk phases, but also balance equations for phase-interfaces and common lines. Furthermore, in this alternative theory, capillary pressure is not a function of saturation only, but also depends on interfacial area. Finally, the difference in macro-scale phase pressures is not equal to capillary pressure if saturations are changing; instead the difference between phase pressures deviates from capillary pressure by a dynamic term.

Until now, there have been only few attempts to use these more physically-based theories and mathematical models for numerical modeling. Nordhaug et al. [2003b] studied two-phase flow including interfacial areas for the case without production of interfacial area (bundle of capillary tube model) and Nordhaug et al. [2003a] used pore-network modeling to obtain estimates of average interfacial velocity. Then, Niessner and Hassanizadeh [2008] modeled two-phase flow including migration and production of interfaces and showed that hysteresis could be reproduced by including interfacial areas and an interfacial area–capillary pressure–saturation surface. This was in line with the conjecture by Hassanizadeh and Gray [1993b] who postulated that capillary hysteresis can be modeled through the inclusion of specific interfacial area in the capillary pressure–saturation relationship.

1.5 Summary

Flow and transport in porous media are universally occuring processes. The understanding of many physical systems and their prediction could be significantly enhanced if current flow and transport models would be advanced, first in terms of their physical correctness and second, once thermodynamically consistent models are developed, in terms of saving computer time and data. Concerning physical correctness, it is well-known that the classical two-phase flow model is not entirely founded on a sound physical basis and therefore, several alternative theories based on thermodynamic principles have been developed. However, apart from very preliminary approaches, no macro-scale numerical models and simulations are based on these theories. Saving computer time and data can be realized using multi-scale multi-physics approaches in time and space which allow to model only the relevant processes and to pick appropriate spatial and temporal scales for the different processes.

The contribution of this book is to close that gap and to

- develop macro-scale models that are based on a thermodynamically consistent set of equations,
- describe two-phase flow in porous media in a more physically-based way by including phase-interfacial areas and dynamics, and to
- model hysteresis as well as kinetic interphase mass and heat transfer based on the knowledge of interfacial area.

As a perspective, it is shown how this new model for two-phase flow can be embedded into a multi-scale multi-physics framework.

1.6 Outline of this thesis

In order to present the current state of two-phase flow modeling in porous media, the classical two-phase flow theory in porous media will be described in detail in Chap. 2. Therefore, basic concepts and parameters, balance equations as well as constitutive relationships and equations of state will be presented. Then, in Chap. 3, the alternative interfacial-area-based approach of Hassanizadeh and Gray [1990] will be introduced and extended for interphase mass and energy transfer. It will be shown how balance equations are formulated for phases and interfaces, how they are upscaled to the macro scale, and how constitutive relationships can be obtained by exploiting the second law of thermodynamics. Next, in Chap. 4, a brief overview of numerical models will be given and the applied numerical scheme will be addressed. In Chap. 5, a number of modeling examples will be shown which illustrate the application of the interfacial-area-based approach to modeling hysteresis and kinetic interphase mass and energy transfer. Chap. 6 embeds the presented equations in a multi-scale multi-physics framework where indicators and dimensionless parameters help to decide whether the more physically correct interfacial-area-based model needs to be solved or whether the classical model gives sufficiently good results. Finally, conclusions will be drawn in Chap. 7.

2 Classical two-phase flow theory

The classical two-phase flow approach postulates conservation equations for mass, momentum, and energy of phases, mostly directly at the scale of interest. The system is then closed by postulating constitutive relationships, also directly at the considered scale.

In this chapter, the scale definition used in this work will be introduced first (Sec. 2.1). Next, in Sec. 2.2, basic fluid and material properties are defined. Then, in Sec. 2.3, balance equations for mass, momentum, and energy will be presented and discussed. In order to close the resulting system of balance equations, constitutive relationships and equations of state are needed. These are addressed in Sec. 2.4. The main issues of the classical model are then summed up in Sec. 2.5. For a comprehensive overview of the classical two-phase flow modeling, we refer to Helmig [1997].

As a starting point for the following discussions, a few basic definitions are needed:

Phase and component A fluid phase is a continuum of fluid which has a sharp interface to other such continua. Physical fluid properties are discontinuous across this interface. In reality, two fluids are always soluble in each other, at least in small amounts. If the solubility of the two fluids in one another is small (e.g. the solubility of water in oil and vice versa) and a sharp interface between the two fluids can be detected, they are considered as two separate phases.

To understand, why this sharp interface between two fluids is formed, one has to look at the molecular scale, see Fig. 2.1 which depicts water as an example. Within the water phase (case A), a water molecule is surrounded by other water molecules. The resulting force of the attracting forces is zero (red arrows). Considering the surface of the water, i.e. the interface between water and gas, one can see that the situation is different. Further water molecules are only found in one hemisphere around

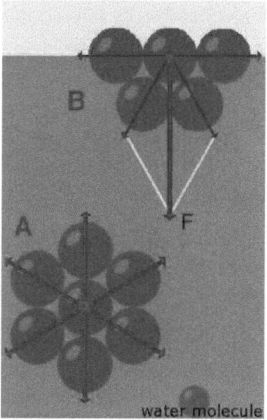

Figure 2.1: Interfacial tension between water phase and gas phase (Source: Fonds der Chemischen Industrie, Germany; image series "Tenside").

a water molecule at the surface (case B), the other hemisphere is occupied by gas. Thus, the resulting force F caused by the dipolar moment of the water molecules points downwards into the water. The result is an interfacial tension between water and gas which in turn leads to a pressure difference across the phase interface.

Note, that gases are always easily miscible with one another and thus, only form one phase. That means, while a system might consist of several liquid phases, there will always be one single gas phase.

Strictly speaking, the solid matrix also forms a phase, the solid phase. However, speaking of phases in this work, it is only referred to fluid phases, i.e. a "two-phase system" represents a system with two fluids which are not soluble in one another and a solid matrix. To denote systems with insoluble (negligibly soluble) phases, the term "multi-phase system" is used; if the phases are soluble in one another, the term "multi-phase multi-component system" is applied.

A phase usually consists of several components considering the fact that in reality, phases are never absolutely insoluble. Components can either be pure chemical substances—elements or molecules, or consist of several substances which form a unit with characteristic physical properties, such as air. Thus, it depends on the model problem which substances or mixtures of substances are considered to be a component. The choice of the components is essential, as balance equations for multi-phase–multi-component systems are in general formulated with respect to components.

Mechanical, chemical, and thermal equilibrium Several types of equilibrium play an important role in the considerations of this work. **Mechanical equilibrium** is fulfilled if all forces acting on a body sum up to zero. Using the usual assumptions of the classical model, mechanical equilibrium directly results from the definition of capillary pressure at the macroscopic scale. **Chemical equilibrium** denotes the fact that the activity of a component is the same in all fluid phases. If the temperatures of all phases (including the solid phase) are the same, the system is in **thermal equilibrium**. If all three equilibrium conditions are fulfilled (mechanical, chemical, and thermal equilibrium), one speaks of **thermodynamical equilibrium**. An important issue is the region where an equilibrium condition is fulfilled. Depending on this region, one may distiguish between different types of equilibria in the following way:

- **global equilibrium** is given if the condition for a certain type of equilibrium is fulfilled within the whole considered system. For example, when considering a fuel cell, global thermal equilibrium would imply that the temperatures of liquid, gas, and solid phase are the same and are constant within the whole fuel cell.

- **local equilibrium** means that a certain type of equilibrium is fulfilled within a defined volume that is smaller than the system size. Usually, the size of this volume is chosen as the size of a so-called representative elementary volume (for a detailed definition of this term, see Sec. 2.1).

- **interface equilibrium** at an interface between phase α and phase β is given if the equilibrium condition is fulfilled only at this $\alpha\beta$-interface.

2.1 Definition of scales

In this section, the definition of scales used in this work is given, being conscious of the fact, that a variety of different definitions is in use.

In theory, all physical processes and properties could be derived from the properties of single molecules and their interactions on a very small scale, the **molecular scale**, see Fig. 2.2, lower left picture. Fluid properties, such as boiling point, density, viscosity, interfacial tensions can be explained by the structure, steric order and dipolar moments of these molecules. However, as one would have to deal with a very large number of molecules (1 g of water contains $3.3 \cdot 10^{22}$ molecules of H_2O), it is not possible to make computations on scales of practical interest for engineering purposes. For this reason, averaging is applied over a large number of molecules

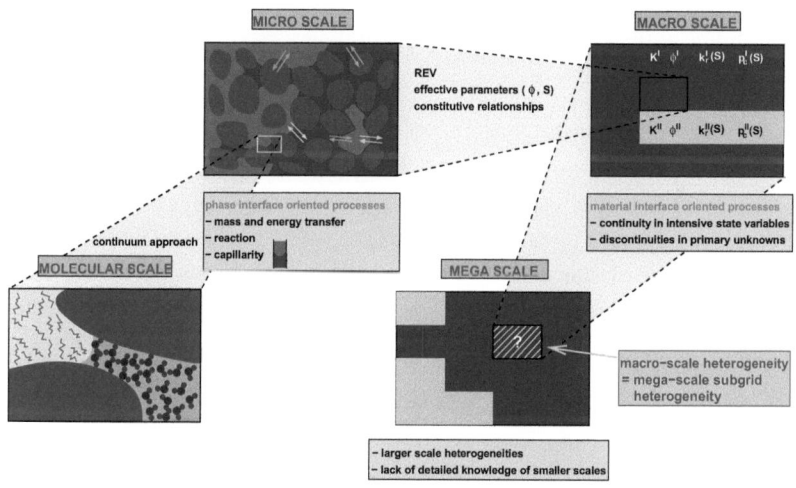

Figure 2.2: Definition of scales.

which is equivalent to assuming matter to be continuous. Single molecules can then no longer be detected, but one can describe substances by new physical parameters representing a large number of molecules, like density (see Sec. 2.2.2), viscosity (Sec. 2.2.3), boiling point etc. This continuity assumption leads to the **micro scale** (also called pore scale). This is the largest scale, where a clear separation of phases can be detected, i.e. one can observe interfaces between the fluid phases as well as fluid–solid interfaces. Therefore, phase-interface oriented processes, such as mass transfer between fluid phases, reaction, energy transfer across fluid–fluid and solid–fluid interfaces, or capillarity should optimally be considered on this scale, otherwise phase interfaces and their interfacial areas will not be correctly resolved.

In fluid mechanics, the Navier-Stokes or Stokes equation is the appropriate description of micro-scale processes.

A further averaging procedure, the averaging of micro-scale properties over a **representative elementary volume** (REV) produces new effective parameters, such as porosity and saturation. Also, so-called constitutive relationships (Sec. 2.4) are needed to bridge the gap from the micro scale to the **macro scale**. Finally, due to the averaging to the macro scale, entirely new equations evolve.

Let us consider one of these parameters, namely the porosity ϕ, which is defined by the volume of the void space in a defined volume element divided by the total volume of this volume element. Then, it becomes obvious, that if a small averaging volume is chosen it might well lie either totally within solid rock or totally within the pore space, i.e. $\phi = 0$ or $\phi = 1$, see Fig. 2.3, left hand side. Enlarging this volume, one will encounter porosity oscillations starting from extreme values and stabilizing at a more or less constant value, until larger scale heterogeneities are included in the averaging volume. The REV is the minimum volume for which the averaging parameters remain constant for the first time when further enlarging this volume, see Fig. 2.3, right hand side. The inclusion of larger scale heterogeneities leads to deviations from this constant value. Having determined the REV size, a further

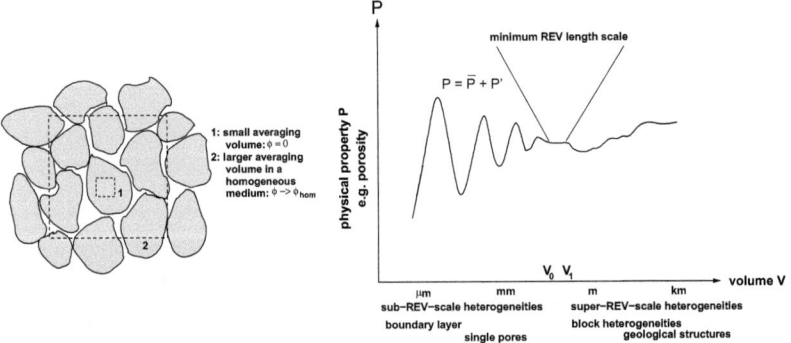

Figure 2.3: Definition of the representative elementary volume (picture on the right hand side after Bear [1972]).

effective parameter can be defined: the saturation of a fluid phase α. This property is the ratio of the volume of phase α within the REV over the total void space within the REV. Thus, it follows automatically that the saturations of all fluid phases within an REV sum up to one,

$$\sum_\alpha S_\alpha = 1. \tag{2.1}$$

Fig. 2.4 illustrates how the indices α are assigned in case of a two-fluid system. The

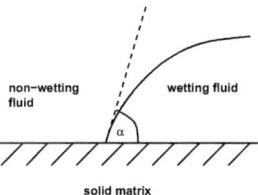

Figure 2.4: Wetting angle α between a wetting and a non-wetting fluid.

fluid with a wetting angle $\alpha < 90°$ is called **wetting fluid** with respect to the solid phase and is denoted by the subscript w, and the fluid with an obtuse wetting angle is called **non-wetting fluid** (subscript n).

In principle, the REV property has to be tested for every single porous medium parameter and additionally, for process parameters. The REV size is then defined by the largest of these averaging sizes. In reality, however, the REV property is often calculated for an easy-to-determine parameter, such as porosity, and assumed to be valid for all processes and parameters.

Due to the averaging process over the REV, discontinuities and interfaces between the fluid phases disappear, but one now has to deal with material discontinuities. These material discontinuities provide new challenges as they may lead to discontinuities in primary variables (e.g. saturation) when assuming continuity of intensive thermodynamical variables (e.g. capillary pressure). The Darcy law is the widely-used model for fluid mechanical problems on the macro scale.

Often, one is interested in larger-scale simulations with high spatial extensions. Therefore, the macro-scale equations are upscaled to an even larger volume-averaged scale, the **mega scale**. On the mega scale, the same effective parameters as on the macro scale occur, but new terms in the equations are needed to account for the subgrid effects coming from the smaller scales.

2.2 Definition of basic fluid and material properties

In this section, all the ingredients of the macro-scale two-phase flow conservation equations will be presented. Therefore, further basic fluid and material properties besides porosity and saturation will be introduced. Specifically, in this section, mass and mole fractions, density, viscosity, intrinsic and relative permeability, capillary pressure, as well as internal energy, enthalpy, heat capacity, and thermal conductivity will be addressed which are of high importance for understanding the two-phase flow balance equations. Parameters will only be introduced in this section. Constitutive relationships and equations of state will be discussed in Sec. 2.4.

2.2.1 Mass and mole fractions, concentrations

For the modeling of two-phase two-component processes, the knowledge about the phase saturations alone is not sufficient as it only gives information about the volume fraction of a fluid within an REV, but not about the composition of the phases that is a result of interphase mass transfer. In case these interphase mass transfer processes are important, measures accounting for the phase composition, that is the amount of components that make up the fluid phases, are needed.

One possibility to describe this composition is to give the dimensionless **mass fraction** X_α^κ of a component κ in phase α. It is defined as the mass of component κ in phase α within an REV over the total mass of phase α within the REV, $X_\alpha^\kappa = \frac{m_\alpha^\kappa}{\sum_\kappa m_\alpha^\kappa}$. Thus, it is clear that

$$\sum_\kappa X_\alpha^\kappa = 1, \qquad (2.2)$$

i.e. the sum of the mass fractions of all components in a phase has to be equal to 1 within an REV.

Analogously to the definition of mass fractions, it is also possible to define **mole fractions** x_α^κ of a component κ in phase α. They describe the number of moles of component κ in phase α, n_α^κ, over the total number of moles n_α of phase α,

$$x_\alpha^\kappa = \frac{n_\alpha^\kappa}{n_\alpha} = \frac{n_\alpha^\kappa}{\sum_\kappa n_\alpha^\kappa}. \qquad (2.3)$$

Like the mass fractions, the mole fractions also sum up to 1, $\sum_\kappa x_\alpha^\kappa = 1$.

Alternatively, it is possible to use the **concentrations** C_α^κ which are defined as the mass of component κ in phase α, m_α^κ within a volume V (usually, an REV) over that volume,

$$C_\alpha^\kappa = \frac{m_\alpha^\kappa}{V}. \qquad (2.4)$$

Unlike mass fractions and mole fractions, these concentrations are not dimensionless, but have the unit $[\text{kg}/\text{m}^3]$.

2.2.2 Density

The macro-scale mass density ρ of a substance is the ratio of the mass m of that substance in a certain volume V (typically, an REV) over that volume:

$$\rho = \frac{m}{V}. \qquad (2.5)$$

Sometimes, molar densities are used instead of mass densities. They are defined as the ratio of the amount of substance n within an REV divided by the volume of an REV,

$$\rho_{mole} = \frac{n}{V}. \qquad (2.6)$$

Mass and molar density are related via $\rho = \rho_{mole} \cdot M$, where M is the molar mass of the substance. Unless otherwise noted, mass densities will be used within this work.

For a fluid phase α, density is in general dependent on the phase pressure p_α and temperature T_α, as well as on the composition x_α^κ of the phase:

$$\rho_\alpha = \rho_\alpha(p_\alpha, T, x_\alpha^\kappa). \qquad (2.7)$$

For details on the equation of state for phase densities, see Sec. 2.4.

2.2.3 Viscosity

Viscosity is a measure for the resistance of a fluid to deformation under shear stress. At solid boundaries that do not move, the so-called no-slip condition applies meaning that the fluid velocity directly at the solid boundary is equal to zero. This implies that a velocity profile develops that is zero at the boundary and increases with increasing distance from the wall. Depending on the fluid, different relations between shear stress τ and velocity gradient $\frac{du}{dy}$ can be established that can all be summarized under the general relationship

$$\tau = \tau_f + f\left(\left(\frac{du}{dy}\right)^n\right). \tag{2.8}$$

Different values of n and τ_f determine different fluid behavior. Among the most often encountered fluids are Newtonian fluids. For this type of fluids, $\tau_f = 0$ and $n = 1$, i.e. the shear stress is proportional to the velocity gradient. Water and air, for example, belong to this class of fluids. The proportionality factor is called *dynamic viscosity* μ (in Pa·s), i.e.,

$$\tau = \mu \frac{du}{dy}. \tag{2.9}$$

Alternatively to the dynamic viscosity μ, sometimes, the kinematic viscosity ν (in m²/s) is used. The two of them are related by the fluid density,

$$\nu = \frac{\mu}{\rho}. \tag{2.10}$$

Like density, viscosity is in general dependent on composition, pressure and temperature. Futher details on these dependencies are discussed in Sec. 2.4.

2.2.4 Intrinsic and relative permeability

Permeability quantifies the resistance to flow, or more precisely, the inverse of this resistance. Depending on whether single-phase flow or two-phase flow conditions prevail, different permeability definitions are commonly used.

Single-phase flow Historically, *Henry Darcy* was the first to introduce a measure of permeability, see Darcy [1856]. He investigated the flow of water through a column filled with a porous material and found out that the water flow rate Q is proportional to the difference in water levels at both sides of the column, Δh,

$$Q \propto \Delta h. \tag{2.11}$$

He called this proportionality factor k_f, hydraulic conductivity, which is measured in m/s. For more details on Darcy's experiment, see Sec. 2.3.2.2.

It could be shown later that in anisotropic porous media, this hydraulic conductivity is in general a tensor, $\underline{\underline{k}}_f$. Also, experiments with fluids different from water were carried through. It could be concluded that the hydraulic conductivity depends on the specific choice of fluid. Once a fluid is chosen and its density and viscosity are known, a fluid-independent parameter accounting for the permeability of the porous medium, $\underline{\underline{K}}$, can be calculated via

$$\underline{\underline{K}} = \underline{\underline{k}}_f \frac{\mu}{\rho g}. \tag{2.12}$$

This fluid-independent permeability measure is called *intrinsic permeability*. It characterizes the porous medium and is of dimensions $[m^2]$.

Two-phase flow In case the pore space is not only occupied by the porous matrix and a single flowing fluid, but by a second fluid, there is no longer a proportionality relation between flow rate of a phase α, Q_α, and hydraulic head gradient Δh_α of that phase. Still, researchers have been trying to somehow maintain this proportionality by postulating that there exists a "proportionality factor" which is called total permeability $\underline{\underline{K}}_{tot,\alpha}$ (in $[m^2]$) and which is dependent on the saturation of phase α. Then, this total permeability is split into a product of the above-defined intrinsic permeability $\underline{\underline{K}}$ and a dimensionless parameter called relative permeability $k_{r\alpha}$,

$$\underline{\underline{K}}_{tot,\alpha} = \underline{\underline{K}} \cdot k_{r\alpha}. \tag{2.13}$$

As intrinsic permeability is a property of the porous matrix only, the saturation dependence is put into relative permeability $k_{r\alpha}$. This functional relationship is determined in a way that it maintains the "proportionality" between flow rate of phase α and hydraulic head gradient of phase α. This "proportionality", however, is only valid for either drainage or imbibition. If both drainage and imbibition occur, the relationship between relative permeability and saturation is not single-valued. Instead, the relative permeability–saturation relationship is hysteretic. For more details on the constitutive relationship between relative permeability and saturation, see Sec. 2.4.

It seems obvious that the attempt to maintain the proportionality between flow rate and head gradient is an ad-hoc procedure. It might be, and it has indeed been shown, that in two-phase flow, there are more driving forces for flow than hydraulic head gradients only. The effect of these missing driving forces is then lumped into relative permeability. This is where the alternative theory that will be introduced in Chap. 3 proposes a way-out in suggesting a more physically-based modeling of two-phase flow by including all relevant driving forces. Only then, possible simplifications of the general equations are considered.

2.2.5 Capillary pressure

For the understanding of macro-scale capillary pressure it is absolutely necessary to start considerations at the pore scale as capillary pressure is an intrinsically pore-scale quantity for which a macro-scale equivalent needs to be sought. The latter is not straight forward. Capillary pressure is an effect that occurs if more than one fluid phase is present. It is a phenomenon that results from the fact that systems tend towards a state of a minimum of free energy. This implies that in free fluids (e.g. water vapor in air) the system tends to minimize the interfacial area between the fluids. In case of water vapor this leads to spherical droplets. In order to increase the interfacial area between two phases α and β by an infinitesimal area element dA, work dW has to be done which can be expressed by

$$dW = \sigma_{\alpha\beta} dA. \qquad (2.14)$$

Here, $\sigma_{\alpha\beta}$ is called the interfacial tension between phase α and phase β and measured in units of energy per area or [N/m].

In a porous medium, the situation is much more complicated due to the presence of the solid phase. Here, additionally, adhesion forces play a role close to the wall. These adhesion forces are generally larger for one of the two fluid phases. The fluid where the adhesion forces are larger is called the wetting fluid which respect to the solid matrix. It can be easily detected by the fact that the contact angle α with the solid phase is smaller than $\frac{\pi}{2}$. Fig. 2.5 shows the situation in an idealized pore throat represented by a capillary tube of diameter d.

The vaulted interface causes a pressure jump across that interface. The pressure is higher on the concave side of the interface (this is obviously the side where the non-wetting phase (n-phase) is located) and lower on the convex side (the side of the

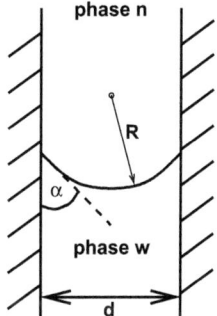

Figure 2.5: Fluid–fluid interface in a capillary tube.

wetting fluid w). The magnitude of the pressure jump can be calculated by

$$p_n - p_w = \frac{2\sigma_{wn}}{R}, \qquad (2.15)$$

where p_n denotes the pressure at the interface on the non-wetting side and p_w is the pressure on the wetting side of the interface directly at the interface. The radius of the curved interface is denoted by R. In case of equilibrium, i.e. if the fluid phases do not move, the pressure jump across a wn-interface is also called *capillary pressure* and denoted by p_c. It is obvious that this pressure jump increases with growing curvature of the interface, i.e. with decreasing R. For special geometries of the pore throats like cylindrical tubes, an easy relation between R, the radius of curvature, and the diameter d of the tube can then be established. In a cylindrical tube, the capillary pressure can thus also be expressed as a function of the diameter of the tube and the wettinge angle α,

$$p_c = \frac{4\sigma \cos \alpha}{d}. \qquad (2.16)$$

This resulting equation is called *Young-Laplace* equation.

Up to this point, all considerations regarding capillary pressure have taken place on the pore scale so far. For macro-scale modeling, a macro-scale version of capillary pressure seems to be inevitable. However, as easy as it may seem at a first glance, this is an absolutely challenging task and still an unsolved research problem (Nordbotten et al. [2008], Korteland et al. [2009]). The most commonly used definition of macro-scale capillary pressure is the intrinsic phase-volume-average of the pore-scale capillary pressure. This means that the macro-scale phase pressure of phase α is obtained by averaging over that volume V_α of the REV where phase α is actually present. Macro-scale capillary pressure is then equal to the difference of macro-scale non-wetting and wetting phase pressure,

$$\bar{p}_c = \bar{p}_n - \bar{p}_w, \qquad (2.17)$$

with

$$\bar{p}_\alpha = \frac{1}{V_\alpha} \int_{V_\alpha} p_\alpha \, dV \qquad (2.18)$$

In the following chapters, the overbars will be left out, tacitly implying that the considered pressures are always macro-scale pressures.

Independently of the averaging procedure, the classical model is a non-closed model problem at first hand. In the classical model, it is generally postulated that a closure of the system is applicable which assumes that macro-scale capillary pressure is a function of wetting-phase saturation only. This constitutive relationship $p_c(S_w)$ will be discussed further in Sec. 2.4.

2.2.6 Internal energy, enthalpy, and heat capacity

The internal energy of a system or phase is the sum of kinetic energy due to the motion of molecules (translational, rotational, vibrational) and the potential energy associated with the vibrational and electric energy of atoms within molecules or crystals. It is usually denoted by U and measured in Joule [J]. More often than the internal energy itself, the specific internal energy is used that is obtained by dividing U by the mass of the system or phase within a representative elementary volume and denoted by u (in [J/kg]).

When heating a system, the amount of energy that is supplied to the system is not equal to the change dU in internal energy of that system if the volume is allowed to change. Instead, the amount of supplied energy is given by the enthalpy change dH which is equal to

$$dH = dU + p\,dV + V\,dp, \qquad (2.19)$$

where dV is the volume change in [m³] and dp is the pressure change in [Pa]. Thus, the difference between internal energy and enthalpy is equal to the sum of the volume changing work at constant pressure (isobaric conditions), $dW_p = p\,dV$ and the work done to change the pressure at constant volume, $dW_V = V dp$. Eq. (2.19) can also be written in terms of specific variables after dividing by the mass of the system or phase within an REV,

$$dh = du + pdv + vdp, \qquad (2.20)$$

where h is the specific enthalpy and $v = \frac{1}{\rho}$ is the specific volume (the inverse of density).

Often, the so-called specific heat capacities are very useful quantities. The specific heat capacity at constant volume, c_v (in $\frac{J}{kg \cdot K}$), is defined by

$$c_V = \left.\frac{\partial u}{\partial T}\right|_V, \qquad (2.21)$$

and the specific heat capacity at constant pressure, c_p (also in $\frac{J}{kg \cdot K}$), by

$$c_p = \left.\frac{\partial h}{\partial T}\right|_p, \qquad (2.22)$$

For ideal gases, the two heat capacities are related by

$$c_p - c_V = \frac{\mathcal{R}}{M}, \qquad (2.23)$$

where $\mathcal{R}$ is the universal gas constant and M is the molar mass.

Both internal energy and enthalpy are functions of pressure and temperature. The relevant equations of state will be discussed in Sec. 2.4.

2.2.7 Thermal conductivity

Thermal conductivity indicates a material's ability to conduct heat. It is denoted by λ and measured in $\frac{W}{mK}$. Its value is generally highest for a solid phase and lowest for gases,

$$\lambda_{solid} > \lambda_{liquid} > \lambda_{gas}. \qquad (2.24)$$

Heat conduction is generally described by *Fourier's law* which describes the diffusive transport of heat due to a temperature gradient and thus, very much ressembles *Fick's law* of diffusion,

$$\underline{F}^{th} = -\lambda \nabla T. \qquad (2.25)$$

Here, $\underline{F}^{th}$ is the conductive heat flux. Depending on the considered material, λ may also be a tensor.

In the classical two-phase flow theory, local thermal equilibrium (see introduction of Chap. 2) is most often assumed. This means, that the temperatures of all phases within an REV are the same. The reason for this assumption is not always the physical justification of this assumption, but often the fact that a physically based description of local thermal non-equilibrium between all three phases is not possible in the classical model. For such a physically based description, phase-interfacial areas would have to be known as the heat transfer takes place across these interfaces and the area of these interfaces thus limits the transfer rates. Thus, in the classical model, it is impossible to account for the highly different thermal conductivities of the solid matrix and possible liquid and gaseous phases. Instead, some way has to be found to calculate an effective thermal conductivity of the fluid-filled porous medium, λ_{pm}. A simple, but often used approach is to weigh the thermal conductivities of the single phases by their volume ratios,

$$\lambda_{pm} = \lambda_s(1-\phi) + \lambda_w \phi S_w + \lambda_n \phi S_n. \tag{2.26}$$

More complex models are available. However, they all only represent empirical approximations of an effective thermal conductivity. The only way to account for the thermal conductivities of the single phases in a physically based way would be to include separate balance equations for the phases and calculate exchange terms proportional to the respective phase-interfacial areas. This is the approach of the alterative and thermodynamically based model that will be followed in Chap. 3.

2.3 Balance equations

In the classical model, balance equations for two-phase flow and transport are often constructed by directly formulating conservation equations on the macro scale. Therefore, most often, an Eulerian framework is chosen where the conserved properties are balanced for a control volume that is fixed in space. This is the approach we will pursue in this work as well. Generally, this control volume coincides with an REV. The properties for which conservation equations are formulated are generally mass, momentum, and energy. Mass conservation is addressed in Sec. 2.3.1, momentum conservation including Darcy's law as a possible simplification is discussed in Sec. 2.3.2, and energy conservation finally is the topic of Sec. 2.3.3.

2.3.1 Mass balance

The mass conservation equation states that the change of mass within an REV is equal to the difference between inflow into and outflow out of the REV plus the difference of sources and sinks. Mass balance equations will be considered separately for two-phase conditions (see Sec. 2.3.1.1) and for two-phase two-component systems (Sec. 2.3.1.2).

2.3.1.1 Two-phase flow

Considering a control volume (i.e. an REV), see Fig. 2.6, a part of the volume is occupied by the solid matrix, a part by the wetting phase and the rest by the non-wetting phase. Balancing the mass within the control volume, the temporal change of mass M_α within one phase is governed by the exchange of mass I_α with the other phase, by sources and sinks Q_α of phase α within the control volume and by the exchange of mass A_α with other control volumes. Thus, a mass balance can be set up which is of the form

$$M_\alpha - A_\alpha - I_\alpha - Q_\alpha = 0, \qquad (2.27)$$

where inflow of mass A_α into the control volume, the flux of mass I_α to phase α, and a source Q_α have positive sign by definition.

Now, the single terms are considered in more detail.

The **storage term** M_α is the temporal change of mass resulting from the integral over a control volume Ω of the phase density ρ_α multiplied by porosity ϕ and the saturation S_α of phase α:

$$M_\alpha = \frac{\partial}{\partial t} \int_\Omega \phi \rho_\alpha S_\alpha \, d\Omega. \qquad (2.28)$$

To determine the **flux term** A_α across the boundary of a control volume, the advective flux across the boundary Γ of the control volume Ω is considered:

$$A_\alpha = \int_\Gamma (\rho_\alpha \underline{v}_\alpha) \cdot \underline{n} \, d\Gamma, \qquad (2.29)$$

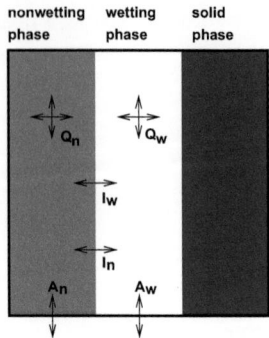

Figure 2.6: Control volume for the phase mass balance.

where $\underline{v}_\alpha$ is the Darcy velocity of phase α which will be discussed later on in this section.

To transform Eq. (2.29), the Gauss theorem

$$\int_\Gamma \underline{f} \cdot \underline{n} \, d\Gamma = \int_\Omega \nabla \cdot \underline{f} \, d\Omega \qquad (2.30)$$

is applied which is valid for any vector-valued function $\underline{f}$. This yields

$$A_\alpha = \int_\Omega \nabla \cdot (\rho_\alpha \underline{v}_\alpha) \, d\Omega. \qquad (2.31)$$

Finally, the **mass exchange term** between fluid phases for multi-phase systems is zero by definition, i.e.

$$I_\alpha = 0. \qquad (2.32)$$

The **source / sink term** is given by

$$Q_\alpha = \int_\Omega (\rho_\alpha q_\alpha) \, d\Omega, \qquad (2.33)$$

where q_α is an external source or sink of phase α (wells, groundwater recharge, etc.), depending on its sign.

Inserting Eq.s (2.28), (2.31), (2.32), and (2.33) into Eq. (2.27) results in

$$\frac{\partial}{\partial t} \int_\Omega (\phi \rho_\alpha S_\alpha) \, d\Omega + \int_\Omega \nabla \cdot (\rho_\alpha \underline{v}_\alpha) \, d\Omega - \int_\Omega (\rho_\alpha q_\alpha) \, d\Omega = 0. \qquad (2.34)$$

Under certain mathematical assumptions, it is possible to write this equation in differential form as

$$\frac{\partial (\phi \rho_\alpha S_\alpha)}{\partial t} + \nabla \cdot (\rho_\alpha \underline{v}_\alpha) - \rho_\alpha q_\alpha = 0. \qquad (2.35)$$

2.3.1.2 Two-phase–two component flow and transport

For the mass balance of a two-phase–two-component system, similar considerations can be made as for the system without mass transfer between phases. Now, one needs to balance over each component κ in each phase α,

$$M_\alpha^\kappa - A_\alpha^\kappa - I^\kappa - Q_\alpha^\kappa = 0, \qquad (2.36)$$

see Fig. 2.7. Note that due to mass conservation $|I^\kappa| := |I_w^\kappa| = |I_n^\kappa|$. The signs are defined analogously to the balance equation for a two-phase system given in Eq. (2.27).

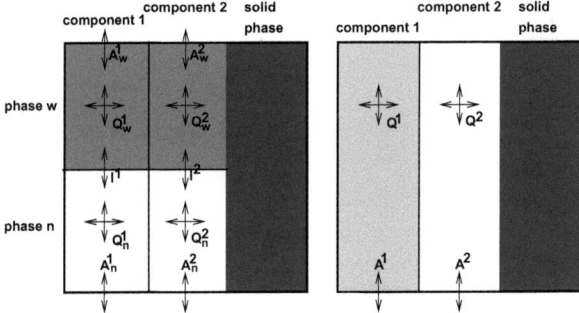

Figure 2.7: Control volume for the component mass balance for each component in each fluid phase (left hand side) and for the total component masses (right hand side).

To set up the mass balance for components, it is possible to either balance over each component in each phase (left hand picture of Fig. 2.7), or to balance over the total mass of a component (right hand picture of Fig. 2.7). The balance over the total masses is advantageous as then only two balance equations instead of $2^2 = 4$ equations are needed and the exchange terms I^κ disappear, similar to the two-phase case. Thus, Eq. (2.37) reduces to

$$M^\kappa - A^\kappa - Q^\kappa = 0 \tag{2.37}$$

As for the two-phase case, the terms M^κ, A^κ, and Q^κ can be obtained as:

$$M^\kappa = \frac{\partial \left(\phi \sum_\alpha C_\alpha^\kappa\right)}{\partial t}, \tag{2.38}$$

$$A^\kappa = \sum_\alpha \nabla \cdot (C_\alpha^\kappa \underline{v}_\alpha + \underline{\underline{D}}_{pm}^\kappa \nabla C_\alpha^\kappa), \text{ and} \tag{2.39}$$

$$Q^\kappa = q^\kappa, \tag{2.40}$$

where q^κ is a source / sink term for component κ and $\underline{\underline{D}}_{pm}^\kappa$ is the tensor of hydrodynamic dispersion of component κ. In this tensor $\underline{\underline{D}}_{pm}^\kappa$, both diffusion and dispersion are subsumed.

Inserting M^κ, A^κ, and Q^κ into Eq. (2.36) results in

$$\frac{\partial C^\kappa}{\partial t} + \sum_\alpha \nabla \cdot (C_\alpha^\kappa \underline{v}_\alpha + \underline{\underline{D}}_{pm}^\kappa \nabla C_\alpha^\kappa) - q^\kappa = 0. \tag{2.41}$$

2.3.2 Momentum balance

The momentum balance can be formulated following exactly the same steps as in Sec. 2.3.1.1 for two-phase flow and in Sec. 2.3.1.2 for two-phase two-component flow, i.e. by identifying the terms M, A, and Q. However, as the parametrization of the momentum balance is much less straightforward than the formulation of the mass balance, considerations are started on the pore scale here in order to allow for a better understanding of the involved processes.

2.3.2.1 General momentum balance

The general momentum balance for phases can be formulated on the pore scale by considering Fig. 2.8 and proceeding analogously to Sec. 2.3.1.1 for the mass balance of two-phase flow on the macro scale. Unlike the mass balance, the momentum

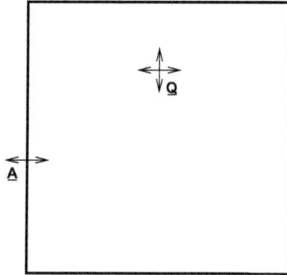

Figure 2.8: Control volume on the pore scale for setting up the pore-scale momentum balance.

balance is a vector-type equation. This means, that the terms $\underline{M}$, $\underline{A}$, and $\underline{Q}$ are vectors and yield,

$$\underline{M} = \frac{\partial (\rho \underline{v})}{\partial t} \tag{2.42}$$

$$\underline{A} = \rho \underline{v} \cdot \nabla \underline{v} \tag{2.43}$$

$$\underline{Q} = \rho^2 \underline{g} - \rho \nabla p + \rho \nabla \cdot \underline{\underline{T}}, \tag{2.44}$$

$$\tag{2.45}$$

where $\underline{\underline{T}}$ is the deviatoric stress tensor. As discussed in Sec. 2.2.3, in case of Newtonian fluids, there is a linear relation between shear stress and velocity gradient which can be used to substitute $\underline{\underline{T}}$ in the term $\underline{Q}$ to obtain

$$\underline{Q} = \rho^2 \underline{g} - \rho \nabla p + \mu \nabla^2 (\rho \underline{v}). \tag{2.46}$$

Note that the term $\underline{Q}$ is not a source / sink term in the classical sense here, but a measure for the external forces which comprise volume and surface forces. Putting the terms $\underline{M}$, $\underline{A}$, and $\underline{Q}$ together yields

$$\frac{\partial (\rho \underline{v})}{\partial t} + \rho \underline{v} \cdot \nabla \underline{v} - \rho^2 \underline{g} + \rho \nabla p - \mu \nabla^2 (\rho \underline{v}) = 0 \tag{2.47}$$

This pore-scale momentum balance equation given by Eq. (2.47) for a Newtonian fluid is called the *Navier-Stokes equation*. Hassanizadeh and Gray [1979a,b] upscaled it by means of volume averaging to the macro scale for the case of incompressible immiscible two-phase flow. Mostly however, in porous media, flow velocities are very slow so that the pore-scale Navier-Stokes equations can be further simplified. The criterion to decide whether this simplification is permitted is whether the flow is "creeping". Creeping flow means that the Reynolds number Re defined as the ratio of inertial forces over viscous forces in a porous medium,

$$\text{Re} = \frac{v \cdot d}{\nu} \qquad (2.48)$$

is less then 1. Here, v is the flow velocity and d should ideally be a characteristic pore size. Mostly however, for practical reasons, a typical grain-size diameter is chosen as this parameter is easier to determine. Eq. (2.48) indicates that creeping flow occurs

- if flow velocity v is small,
- if the characteristic pore size (or grain size) d is small, and
- if the kinematic viscosity ν is high.

If $\text{Re} < 1$ is valid, then inertial forces can be neglected. The Navier-Stokes equations given in Eq. (2.47) then simplify to the so-called Stokes equation,

$$\rho^2 \underline{g} + \rho \nabla p - \mu \nabla^2 (\rho \underline{v}) = 0 \qquad (2.49)$$

In order to use this momentum balance in macro-scale simulations, it needs to be upscaled. Bear and Bachmat [1990], Auriault [2005] showed that under certain conditons, this upscaling procedure yields the famous and widely-used Darcy's law, see Sec. 2.3.2.2.

2.3.2.2 Darcy's law and extensions

Darcy's law is an empirical functional relationship that has been obtained experimentally by *Henry Darcy* (Darcy [1856]) for isothermal water flow under single-phase conditions through an isotropic and homogeneous porous medium in a one-dimensional column. Since then, it has been shown that the law that he found can also be derived more rigorously by homogenization of the Stokes equation under certain assumptions (see e.g. Bear and Bachmat [1990], Auriault [2005]).

Henry Darcy, at the time he developed his famous law, was an engineer of the city of Dijon in France where he investigated the connection between the flow of water in a

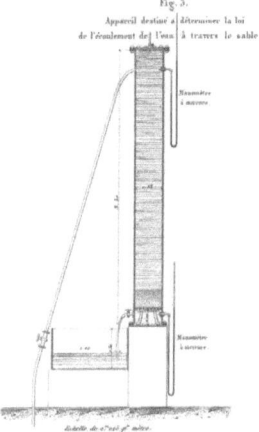

Figure 2.9: Setup of Henry Darcy's original experiment (taken from Darcy [1856]).

sand column to the city's fountains. He studied an experimental setup as is shown in Fig. 2.9 and varied the length and diameter of the column, the porous medium inside the column, as well as the water levels of the inlet and outlet reservoir. From these experiments he could conclude that the flow rate of water through the column, Q, is

- proportional to the cross-sectional area A of the column,
- proportional to the difference in water levels, Δh of the inlet and outlet reservoir, and
- inversely proportional to the length L of the column.

Putting this together, he obtained the famous Darcy's law,

$$Q = k_f A \frac{\Delta h}{L}, \tag{2.50}$$

where k_f is a proportionality constant called hydraulic conductivity. By shrinking the length L it has been postulated that this equation is also valid in differential form,

$$Q = A \cdot v, \tag{2.51}$$

with
$$v = -k_f \frac{\partial h}{\partial x}. \qquad (2.52)$$
where v is called specific discharge or Darcy velocity.

The same functional relationship between hydraulic head gradient and flow rate has been found rigorously by homogenizing the Navier-Stokes equations (Auriault [2005], Bear and Bachmat [1990]), but making the assumptions that inertial forces and friction within the water can be neglected.

Since the development of Darcy's law in 1856, various extensions to the classical Darcy equation have been made. This means that an equation developed for a physically simpler situation has been generalized instead of starting of with a general pore-scale equation, averaging and simplifying it. Specifically, Darcy's law has been extended to (a) non-creeping flow, (b) multi-dimensional domains, (c) non-isotropic inhomogeneous porous media, (d) compressible compositional non-isothermal multi-phase flow, and (e) high-porosity flow. These different extensions will be briefly discussed in the following.

(a) In case that the condition of small Reynolds numbers $\text{Re} < 1$ is not fulfilled, flow is not creeping and on the pore scale, the Navier-Stokes equation instead of the Stokes equation would need to be employed. On the macro scale, this situation has been handled by extending Darcy's law by a quadratic term to the so-called *Forchheimer equation*,
$$v + \beta \rho v^2 = -k_f \frac{\partial h}{\partial x}, \qquad (2.53)$$
where β is a factor that is usually obtained experimentally. Sometimes, even third order terms in velocity are used leading to
$$v + \beta \rho v^2 + \gamma \rho v^3 = -k_f \frac{\partial h}{\partial x}, \qquad (2.54)$$

(b) In the multi-dimensional case, the one-dimensional Darcy equation (2.52) is transformed to a vector-valued equation by replacing the scalar v by a vector and by replacing the partial derivative of h with respect to x by a gradient of h,
$$\underline{v} = -k_f \nabla h. \qquad (2.55)$$

(c) Non-isotropic conditions are handled by replacing the scalar k_f by a tensor $\underline{\underline{k}}_f$. In the very general case, $\underline{\underline{k}}_f$ is a full tensor. However, in many cases, it is possible to apply a principle axis transformation in order to align the coordinate system along the principle directions of permeability and to make $\underline{\underline{k}}_f$ a diagonal matrix. Inhomogeneity is realized in Darcy's law by having a space-dependent k_f value, i.e., $k_f = k_f(\underline{x})$ or $\underline{\underline{k}}_f = \underline{\underline{k}}_f(\underline{x})$.

(d) Compressible flow, compositional flow, multi-phase flow, as well as non-isothermal flow are commonly handled in a relatively easy way: by making parameters in Darcy's law pressure-, component-, phase-, and / or temperature-dependent,

$$v = -k_f\left(p_\alpha, X_\alpha^\kappa, S_\alpha, T\right) \frac{\partial h\left(p_\alpha, X_\alpha^\kappa, S_\alpha, T\right)}{\partial x}. \tag{2.56}$$

(e) In case of porosities close to 1, Brinkman [1947] suggested the following extension to Darcy's law:

$$v = -k_f \frac{\partial h}{\partial x} + \delta \nabla^2 v, \tag{2.57}$$

where δ is called the effective viscosity. Eq. (2.57) is generally called the *Brinkman equation*.

The above extensions of Darcy's law have been combined in various ways yielding macro-scale momentum balances for almost arbitrary porous media applications. Unfortunately, some of these extensions have only been made empirically and are lacking a sound physical basis which leads to a list of problematic issues of the classical two-phase flow approach, see Sec. 3.1.1. What could alternatively be done is to formulate the appropriate pore-scale balance equations involving all occurring processes and average these equations to the macroscopic scale.

2.3.3 Energy balance

In the classical two-phase flow theory, local thermal equilibrium is generally assumed, see introduction of Chap. 2. This assumption implies that the temperatures of all three phase (the two fluid phases and the solid phase) within an REV are the same, $T_w = T_n = T_s$. This means that it is sufficient to formulate a single energy balance for the sum of the thermal energies of all three phases.

Using the previously introduced notation, the terms M, A, I, and Q can be easily obtained. Note that similarly to the mass balance of two-phase two-component flow and transport, the term A consists of an advective and a diffusive part

$$M = \frac{\partial (\phi \rho_w u_w S_w)}{\partial t} + \frac{\partial (\phi \rho_n u_n S_n)}{\partial t} + \frac{\partial ((1-\phi)\rho_s c_s T)}{\partial t} \tag{2.58}$$

$$A = \nabla \cdot (\rho_w h_w \underline{v}_w) + \nabla \cdot (\rho_n h_n \underline{v}_n) - \nabla \cdot (\lambda_{pm} \nabla T) \tag{2.59}$$

$$I = 0 \tag{2.60}$$

$$Q = Q^h \tag{2.61}$$

Putting these terms together and writing the terms for the fluid phases as sums, yields the energy balance equation of the classical two-phase flow theory,

$$\frac{\partial (\phi \sum_\alpha \rho_\alpha u_\alpha S_\alpha)}{\partial t} + \frac{\partial ((1-\phi)\rho_s c_s T)}{\partial t} + \nabla \cdot \left(\sum_\alpha \rho_\alpha h_\alpha \underline{v}_\alpha \right) - \nabla \cdot (\lambda_{pm} \nabla T) - Q^h = 0. \tag{2.62}$$

Recently, first extensions of this model were made by Crone et al. [2002] allowing to partly account for non-equilibrium situations. Specifically, in this model, the two fluid phases are summarized to an entity, the fluid mixture, and the solid phase is considered a separate entity. This means that two energy balances are formulated, one for the fluid mixture, and one for the solid phase. With this approach, non-equilibrium between the fluid mixture and the solid phase can be represented, but it is impossible to account for non-equilibrium between the fluid phases.

Using this conceptual model, Crone et al. [2002] obtained two one-dimensional energy balance equations, the first one for the fluid mixture, and the second one for the solid phase:

$$\frac{\partial (\phi u \rho)}{\partial t} + \frac{\phi \partial (\rho h v)}{\partial x} = \left[\frac{\partial}{\partial x} \left(\lambda_{pm} \frac{\partial T_f}{\partial x} \right) + \alpha a (T_{s,0} - T_f) \right] \tag{2.63}$$

$$\frac{\partial ((1-\phi)\rho_s c_s T_s)}{\partial t} = [\alpha a (T_f - T_{s,0})], \tag{2.64}$$

where α is a heat transfer coefficient and a is the specific surface area of the solid matrix. In this model, mixture density ρ, specific mixture enthalpy h, and specific mixture velocity v, respectively, are given by

$$\rho = S_w \rho_w + S_n \rho_n \tag{2.65}$$

$$h = \frac{1}{\rho} \left(S_w \rho_w h_w + S_n \rho_n h_n \right) \tag{2.66}$$

$$v = \frac{1}{\phi \rho h} \sum_\alpha \left(\rho_\alpha v_\alpha h_\alpha \right). \tag{2.67}$$

The value of $T_{s,0}$ (the temperature at the surface of the grains) is either equal to T_s if heat conduction within the soil grains is neglected (e.g. if the grains are small). Otherwise, it can be estimated by assuming that the grains are sphere-shaped and of equal radius.

Using this procedure, it is possible to account for non-equilibrium effects between the fluid mixture and the solid surface in case the solid surface area is known. However, the model of Crone et al. [2002] cannot take non-equilibrium between the fluid phases into account; also, it contains no macro-scale heat conduction within the solid phase. This means that it tends to overestimate non-equilibrium effects.

2.4 Constitutive relationships and equations of state

A constitutive relationship is a relation between two or more physical quantities which is specific to a material and approximates the response of the material to exernal forces. An equation of state, contrarily, is a relation between state variables, such as temperature, pressure, volume, or internal energy. In this section, equations of state for density, viscosity, internal energy and enthalpy, and the equilibrium composition will be discussed. Also, constitutive relationships for relative permeability and capillary pressure will be addressed.

2.4.1 Density

As mentioned in Sec. 2.2.2, the macro-scale density of a substance is in general a function of pressure and temperature, $\rho = \rho(p, T)$. Thus,

$$d\rho = \frac{\partial \rho}{\partial p} dp + \frac{\partial \rho}{\partial T} dT = \rho \left(\beta_p dp + \beta_T dT \right), \tag{2.68}$$

with $\beta_p = \frac{1}{\rho}\frac{\partial \rho}{\partial p}$ and $\beta_T = \frac{1}{\rho}\frac{\partial \rho}{\partial T}$. For **ideal gases**, the functions β_p and β_T can be calculated exactly. Dividing the ideal gas law,

$$pV = n\mathcal{R}T, \tag{2.69}$$

by the volume V and resolving for $\rho_{mole} = \frac{n}{V}$ leads to

$$\rho_{mole} = \frac{p}{\mathcal{R}T} \tag{2.70}$$

or

$$\rho = \frac{p}{RT}, \tag{2.71}$$

where R is the individual gas constant of a substance. From Eq. (2.71) it directly follows that $\beta_p = \frac{1}{\rho}\frac{1}{RT}$ and $\beta_T = -\frac{1}{\rho}\frac{p}{RT^2}$. Fortunately, ideal gas behavior can be assumed for all gases far below the critical point. Thus, it holds for many gases and vapors at atmospheric conditions. If the assumption of an ideal gas is not valid (**real gas**), deviations will occur and empirical equations are typically employed for the estimation of gas densities. The same is valid for **liquid densities**. There, often, densities are related to temperature and / or pressure at the critical point and a number of empirical fitting parameters.

The **solid matrix** can be assumed to be rigid for many applications, its density is then constant. If the solid matrix is deformable empirical equations are also needed to describe the compressibility.

2.4.2 Viscosity

Viscosity is often highly temperature dependent, but hardly dependent on pressure. The temperature dependence of gases and liquids shows the opposite behavior: while the viscosity of gases increases with increasing temperature the viscosity of liquids decreases with increasing temperature.

For both gases and liquids, extensive tables are available that give viscosity values at different temperatures. Often, formulas are employed that are based on a number of fitting constants. For gases, viscosity also depends strongly on composition. Simple approaches just weigh the dynamic viscosities of the components within the gas phase by their mole fraction,

$$\mu_g = \sum_\kappa \mu_g^\kappa x_g^\kappa. \qquad (2.72)$$

In case of a binary mixture, more sophisticated approaches are available that are e.g. based on the kinetic gas theory.

2.4.3 Internal energy and enthalpy

Internal energy of gases and liquids is generally a function of pressure, temperature, and composition. For water and water vapor, various tables are available in the literature that quantify the dependence of internal energy and enthalpy on pressure and temperature, see e.g. IAPWS (The International Association for the Properties of Water and Steam) [2003]. For water, the mass fractions of other components besides water are often negligibly small (unless e.g. salt water is considered) so that their influence is of minor impact. For gases, the situation is different: their composition may generally vary a lot, such that the composition needs to be accounted for. The simplest way to do that is to proceed analogously to viscosities (see Sec. 2.4.2) and weigh gas enthalpies by their mass fractions,

$$h_g = \sum_\kappa h_g^\kappa X_g^\kappa. \qquad (2.73)$$

Dependencies of specific internal energies on pressure and temperature can be deduced by combining Eq.s (2.21) through (2.23) and discretizing the occuring partial derivatives.

2.4.4 Equilibrium composition

The classical two-phase flow approach usually assumes local chemical equilibrium. This implies that the mass transfer rate accounting for transfer of components from phase α to phase β is the same as the transfer rate from phase β to phase α. Then, the considered two-phase flow system is in local equilibrium with respect to mass transfer and the net interphase exchange rates are zero.

In order to quantitatively determine equilibrium composition, different mathematical approaches are available. In this section, a short summary is given on the physical relationships which govern the local chemical equilibrium state between fluid phases. While in many real-life systems kinetic interphase mass transfer might play an important role, it is a common assumption in hydraulic engineering to postulate chemical equilibrium, if flow velocities are slow in comparison to kinetic processes. More precisely, the characteristic time for mass transfer t_{mt}^κ of a component κ is compared to the characteristic time of advective flow t_a^κ for a component κ,

$$\mathrm{Da}^\kappa = \frac{t_a^\kappa}{t_{mt}^\kappa}, \tag{2.74}$$

where Da^κ is the Damköhler number for mass transfer of component κ. If Da^κ is small, then the system "has no time" to establish local chemical equilibrium as flow rates are so high that chemical equilibrium cannot be reached across phase-interfaces within the characteristic time of advection. If, contrarily, Da^κ is very large, this may indicate that the characteristic time for mass transfer is so small that equilibrium is also not established. For moderate Da^κ, local chemical equilibrium represents the system state in good approximation. In that case, simple equilibrium relationships can be employed to determine the relationship between concentrations in the different phases. While a variety of other equilibrium relationships can be found in the literature, the following discussion will focus on Dalton's law, Raoult's law, and Henry's law.

Dalton's Law

The English scientist John Dalton studied the properties of gas mixtures and stated the following law, known as Dalton's Law:
The total pressure of a gas mixture equals the sum of the pressures of the gases that make up the mixture,

$$p_g = \sum_\kappa p_g^\kappa, \tag{2.75}$$

where p^κ is the pressure of a single component κ ("partial pressure"). The partial pressure p_g^κ is by definition the product of the mole fraction of the respective component in the gas phase and the total pressure of the gas phase,

$$p_g^\kappa = x_g^\kappa p_g. \tag{2.76}$$

Raoult's Law

Raoults law describes the lowering of the vapor pressure of a pure substance in a solution. It relates the vapor pressure of components to the composition of the solution under the simplifying assumption of an ideal solution. The relationship can be derived from the equality of fugacities, see Prausnitz et al. [1967]. According to Raoult's law, the vapor pressure of a solution of component κ is equal to the vapor pressure of the pure substance times the mole fraction of component κ in phase α.

$$p_g^\kappa = x_\alpha^\kappa \cdot p_{vap}^\kappa \tag{2.77}$$

Here, p_{vap}^κ denotes the vapor pressure of pure component κ which is generally a function of temperature.

Henry's Law

Henry's law is valid for ideally diluted solutions and ideal gases. It is especially used for the calculation of the solution of gaseous components in liquids. Considering a system with gaseous component κ, a linear relationship between the mole fraction x_α^κ of component κ in the liquid phase and the partial pressure p_g^κ of κ in the gas phase is obtained,

$$x_\alpha^\kappa = H_{w-n}^\kappa \cdot p_g^\kappa. \tag{2.78}$$

The parameter H_{w-n}^κ denotes the Henry coefficient of component κ with respect to phases w and n which is dependent on temperature.

Fig. 2.10 shows the range of applicability of both Henry's law and Raoult's law for a binary system, where component 1 is a component forming a liquid phase, e.g. water, and component 2 is a component forming the gaseous phase, e.g. air. One can see, that for low mole fractions of component 1 in the system (small amounts of liquid in the gas phase), Henry's law can be applied whereas for mole fractions of component 1 close to 1 (small amounts of component 2 in the liquid phase), Raoult's law is the appropriate description. In general, the solvent follows Raoult's law as it is present in excess, whereas the dissolved substance follows Henry's law as it is highly diluted.

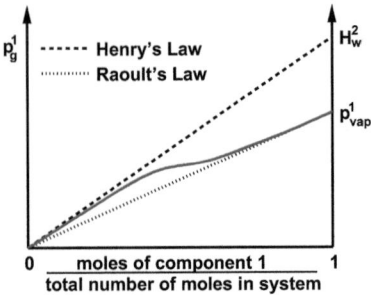

Figure 2.10: Applicability of Henry's law and Raoult's law for a binary gas–liquid system (after Lüdecke and Lüdecke [2000]).

2.4.5 Relative permeability

Relative permeability was introduced in Sec. 2.2.4 as an empirical measure for the increase of resistance against the flow of a fluid phase α due to the presence of a second fluid phase β.

For the determination of relative permeability, a one-dimensional (often horizontal) column or flow cell is filled by a porous medium. Then, either both phases are co-injected in a way that a constant saturation distribution is obtained along the column or a flow rate of a phase is imposed while the other phase is held at constant saturation and the resulting pressure gradient across the column is measured. In both cases, relative permeability can then be obtained by,

$$k_{r\alpha} = \frac{Q_\alpha/A \cdot \mu_\alpha}{K \cdot \Delta p_\alpha/L}, \qquad (2.79)$$

where L is the length of the column and A is its cross sectional area. Intrinsic permeability K and dynamic viscosity μ_α have been introduced previously, in Sec. 2.2.4 and 2.2.3, respectively. For details on the origin of Eq. (2.79), see Sec. 2.3.2.2. Relative permeability values usually range from 0 to 1, but values as high as 16 have been reported, see Berg et al. [2008] and references therein. A relative permeability of phase α of 0 is encountered if phase α is not mobile, i.e. if its saturation is equal to or below the residual saturation. A value of 1, contrarily, occurs for phase α if the α-phase saturation is equal to 1. From these considerations, the conclusion in the classical approach is that relative permeability is a function of saturation. In order to determine the constitutive dependency of the α-phase relative permeability on the saturation of phase α, experiments as shortly outlined above are carried out: initially, the porous medium is fully saturated with wetting phase. Then, both phases are injected co-currently at well-defined flow rates Q_α until steady state is reached. Then, the pressure drop along the column is recorded along with saturations. Next, the fractional flow rate of non-wetting phase is increased stepwise and at each steady state, pressure drop and saturations are recorded. After applying Eq. (2.79), relative permeabilities can be plotted versus saturations and a functional relationship can be fitted.

These fitting functions are often obtained from theoretical considerations based on bundle of capillary tube models. A very popular of these fitting functions, the *Brooks–Corey* model (see Brooks and Corey [1964]), is combined with the bundle of capillary tube model of *Burdine* (see Burdine [1953]) who represents a porous medium by a bundle of tubes with different diameters. Using that conceptual model, Brooks & Corey obtained the following relative permeability functions based on their earlier-developed capillary pressure–saturation function $p_c(S_w)$ (see Sec. 2.4.6),

$$k_{rw} = S_e^{\frac{2+3\lambda}{\lambda}}, \tag{2.80}$$

$$k_{rn} = (1 - S_e)^2 \left(1 - S_e^{\frac{2+\lambda}{\lambda}}\right), \tag{2.81}$$

where λ is a dimensionless parameter characterizing the uniformity of pore sizes and S_e is the so-called effective saturation which is defined by

$$S_e = \frac{S_w - S_{wr}}{1 - S_{wr}}. \tag{2.82}$$

The higher the λ-parameter, the more uniform is the porous medium.

A slightly more complex conceptual model of a porous medium is given by the model of Mualem [1976] who allowed pore diameters to vary also along a single tube. Using *Mualem's* conceptual model, van Genuchten [1980] obtained a capillary pressure–saturation relationship. Based on that, he obtained the following relative permeability–saturation relationships:

$$k_{rw} = \sqrt{S_e}\left[1 - \left(1 - S_e^{\frac{1}{m}}\right)^m\right]^2, \tag{2.83}$$

$$k_{rn} = (1 - S_e)^{\frac{1}{3}}\left(1 - S_e^{\frac{1}{m}}\right)^{2m}. \tag{2.84}$$

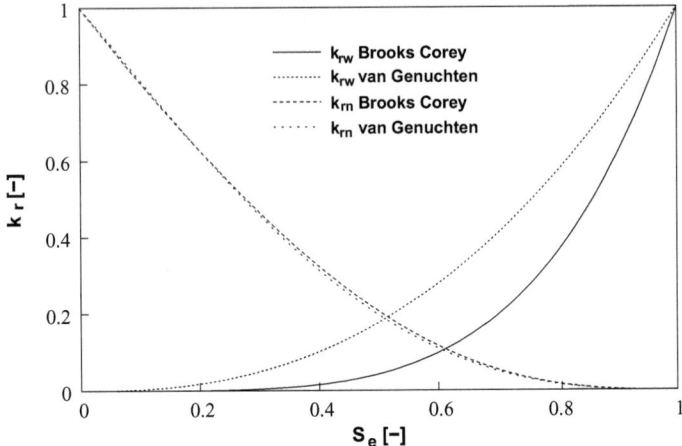

Figure 2.11: Relative permeability–saturation relationships according to the Brooks & Corey and the van Genuchten model.

Here, the fitting parameter is m.

Despite of the fact that these relationships have proven useful for the modeling of two-phase flow, the introduction of relative permeabilities is empirical and only motivated by the ambition to maintain a linear relationship between flow rate and hydraulic head gradient, although this relationship might as well be non-linear due to further driving forces in two-phase flow besides hydraulic head gradients. Typical $k_{r\alpha}(S_e)$ relationships for both Brooks-Corey and van Genuchten model are plotted in Fig. 2.11.

Also, it turned out that the relative permeability–saturation relationships are hysteretic meaning that the functions are not single valued, but that instead, relative permeabilities are history-dependent. As the phenomenon of hysteresis is more pronounced for capillary pressure, it will be discussed in detail in the next section, but it is also an important issue for relative permeabilities.

2.4.6 Capillary pressure

The physics of capillary pressure on the pore scale has been discussed in Sec. 2.2.5. There, capillary pressure is defined as a parameter that was related to interfaces and only dependent on interface properties. Specifically, it was dependent on interfacial tension, wetting angle, and the radius of the capillary tube or pore. At equilibirum, it is exactly equal to the pressure jump across the interface between two fluid phases. On the macro scale, the definition of capillary pressure is much more difficult. It should be a variable that in some way accounts for the effect of pore-scale capillary pressure. However, it obviously has to be defined in a different way as pressure values are only available for volumes (to be exact, for the representative elementary volumes), but not at interfaces.

Therefore, what is commonly done for the definition of macro-scale capillary pressure, is to define it in terms of formulas in the same way as pore-scale capillary pressure,

$$p_c = p_n - p_w, \qquad (2.85)$$

where p_n and p_w now are intrinsic phase-volume-averages of pressure within an REV. Eq. (2.85) indicates, that this macro-scale "capillary pressure" is not related to any pressure jump across phase-interfaces any more, but related to bulk phase pressures. As it has been shown that the macro-scale set of balance equations (mass, momentum, and energy balance) is not closed a priori, but that instead, one equation is missing, a constitutive relationship for the macro-scale capillary pressure is commonly introduced. As capillary pressure obviously changes when wetting-phase saturation changes, a constitutive relationship $p_c = p_c(S_w)$ is introduced. Obviously, for both extreme cases, i.e. $S_w = 0$ and $S_w = 1$, capillary pressure is not defined as phase-interfaces are non-existent and no pressure difference between phases can be defined. In between these extreme saturations, naturally, the wetting phase will preferentially wet the solid surface of the porous matrix and therefore, occupy the smaller pores. This leads to the fact that capillary pressure increases with decreasing wetting-phase saturation.

Once the postulation of p_c depending on S_w was made there was growing interest of experimentalists in measuring capillary pressure–saturation functions. Two basically different measurement types have been established since then. Both of them use a small measurement cell filled with a homogeneous porous medium:

1. In the first measurement type, a non-wetting phase reservoir is connected to one side of the cell and the non-wetting phase pressure is fixed as one boundary condition. On the opposite side, the wetting-phase pressure is fixed. The remaining boundaries are impermeable to flow. Then, the difference between

pressure at the non-wetting reservoir side and at the wetting-reservoir side is increased stepwise (most often, non-wetting phase pressure is increased). After each increase, one needs to wait for the system to reach static conditions. Once the static condition is reached, saturation is determined, either by weighing the sample or by measuring the volume of the outflow This yields one point of the capillary pressure–saturation function. This procedure is repeated to cover the whole range of saturations between zero and one.

2. The second measurement type works basically the other way around: starting with a fully wetting-phase saturated sample, non-wetting phase is stepwise let through a valve into the sample. At two opposite sides of the sample, non-wetting and wetting phase pressure transducers, respectively, are installed. As in the first alternative, at each saturation step, the system equilibrates. Once the fluids have redistributed and come to a rest, the difference between non-wetting and wetting phase pressure is measured across the domain. Each static pair of saturation and capillary pressure yields a point on the capillary pressure–saturation function.

Similarly to relative permeability, conceptual models for capillary pressure–saturation relationships have been derived from bundle of capillary tube models (approach of Mualem [1976] and Burdine [1953]). Based on these models, several parameterizations of capillary pressure–saturation relationships have been postulated. The two most well-known models are the parameterization of van Genuchten [1980],

$$p_c(S_e) = \frac{1}{\alpha} \left(S_e^{-\frac{1}{m}} - 1 \right)^{\frac{1}{m}} \tag{2.86}$$

with the previously introduced parameter m and an additional parameter α as well as the model of Brooks and Corey [1964],

$$p_c = p_d S_e^{-\frac{1}{\lambda}}. \tag{2.87}$$

While for relative permeability, there was no major conceptual difference in the parameterizations of Brooks & Corey and van Genuchten the behavior of the capillary pressure–saturation functions close to a wetting-phase saturation of one is fundamentally different, see Fig. 2.12. Note that for a wetting-phase saturation of 1 (as stated above) capillary pressure is not defined. Still, both models yield a capillary pressure value for this saturation. The assumption of this value (which naturally

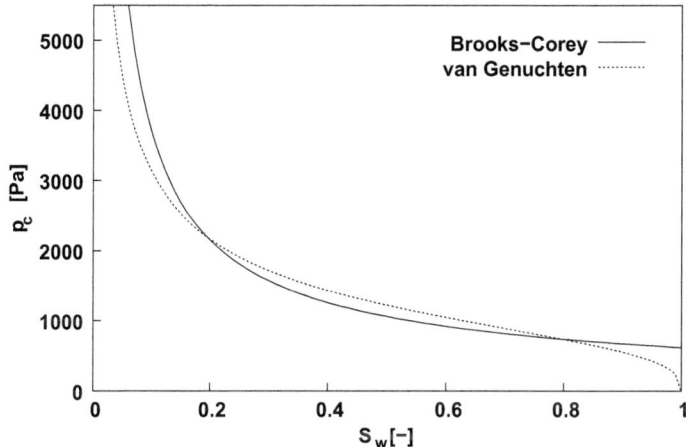

Figure 2.12: Capillary pressure–saturation relationships according to the Brooks & Corey and the van Genuchten model.

highly influences the behavior of the capillary pressure–saturation function for water saturations close to one) makes the difference: while the Brooks & Corey model introduces a macro-scale "entry pressure" which is meant to represent a macro-scale equivalent to the entry pressure given by the Young-Laplace equation on the pore scale, the van Genuchten model assumes that capillary pressure is zero for a water saturation of one. Experimental data as well as data from pore-network models usually show a behavior that lies in between the behavior predicted by the two models, i.e. measured curves usually have a steep increase in capillary pressure close to a water saturation of one, but there is no point of $S_w = 1$, $p_c = 0$.

Unfortunately, measurements have shown that the relationship between capillary pressure and saturation is not unique. Instead, depending on the history of the system and on the process (increasing wetting-phase or increasing non-wetting phase saturation) different capillary pressure–saturation functions are obtained. In fact, it turns out that an infinite number of capillary pressure–saturation functions is possible. However, this family of functions is usually bounded by an upper limiting function and a lower limiting function. The history dependence of the capillary pressure–saturation relationship is called *capillary hysteresis*.

In the following, this phenomenon will be studied and illustrated in more detail for a porous medium that is initially fully wetting-phase saturated. All points in the

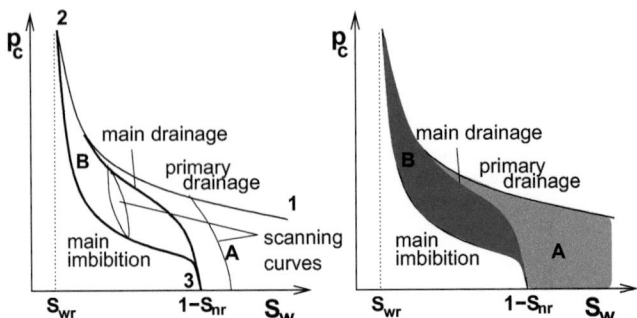

Figure 2.13: Capillary hysteresis for a porous medium that is initially fully wetting-phase saturated. Left hand side: bounding and scanning curves; right hand side: possible capillary pressure–saturation values.

capillary pressure–saturation plane that are discussed here are indicated in Fig. 2.13.

The starting point is indicated by 1 in the left hand picture of Fig. 2.13. Next, non-wetting phase saturation is increased in small steps (each step is an equilibrium stage). Then, capillary pressure follows the so-called *primary drainage curve*. Even if capillary pressure is increased up to a very high value, there will be some water remaining in the system. The volume percentage of this water is called residual (or irreducible) wetting-phase saturation S_{wr}. If the system is drained as far as possible down to a wetting-phase saturation of S_{wr} (point 2) and then imbibed again, capillary pressure follows the *main imbibition curve*. If the system undergoes imbibition as far as possible, at some point, capillary pressure is almost zero and the wetting-phase saturation is at $1 - S_{nr}$ where S_{nr} is the residual (irreducible) non-wetting phase saturation (point 3). Another drainage will produce the main drainage curve. The main drainage curve meets primary drainage and main imbibition curve in point 2. Main drainage and main imbibition curve mark the hysteresis loop: all points on and in between these bounding curve belong to the hysteresis loop. Once the system has undergone a full main drainage and main imbibition cycle capillary pressure–saturation pairs cannot "leave" the hysteresis loop any more. However, if drainage or imbibition is stopped at some point between point 2 and 3 and the process is turned around (switched from drainage to imbibition or vice versa), then, at this intermediate point, so-called *scanning curves* will occur. If one considers the fact that it is possible to stop at any intermediate point on main drainage and main imbibition curve between points 2 and 3 and change the process (drainage to imbibition or vice versa), it is possible to obtain scanning curves at every point within the hysteresis loop (denoted by B in the left hand picture). Repeating the construction of scanning curves, one ends up with area B in the right-hand picture of Fig. 2.13: all p_c–S_w values between main drainage and main imbibition curve (within the hysteresis loop) are possible. Of course, one could also stop the drainage process while still on the primary drainage curve and switch to imbibition: this may lead to the right scanning curve starting on the primary drainage curve in the left-hand picture. Naturally, there is also a whole family of such curves. Their entity is marked by the area A in the right-hand picture. stay within the bounding curves.

Fig. 2.13 suggests that the assumption of capillary pressure being a function of saturation only is incomplete. If it was a function of saturation only, there should be a single-valued capillary pressure–saturation relationship. The fact that there is an area between two bounding curves suggests that capillary pressure be a function of saturation and at least one more variable. In fact, the alternative theory introduces an additional dependency of capillary pressure on specific interfacial areas which indeed has been shown to reduce capillary hysteresis to within the measurement error, see Chap. 3.

An important fact that has been mentioned, but not thoroughly discussed so far is that capillary pressure–saturation relationships are by definition obtained under static conditions. This means that every single point on a capillary pressure–saturation relationship represents an equilibrium state which can be obtained from $p_n - p_w$ under no-flow conditions. However, for porous medium applications, the situation of flowing fluid phases is usually of interest and not the static state of the system after redistribution of fluid phases under no-flow conditions. In case of very small flow velocities, or more precisely, very small time rates of change of saturation $\frac{\partial S_w}{\partial t}$, it may be justified to approximate capillary pressure as $p_c = p_n - p_w$. Contrarily, if the time rate of change of saturation is high, then deviations from the static behavior will occur. In case dynamic effects are important, it is obvious that the quantity measured in experiments is $p_n - p_w$ which is different from p_c. Therefore, it will be discussed about "$p_n - p_w$–saturation relationships" in the following. Capillary pressure–saturation relationships remain uninfluenced by dynamic effects; the important point is that they can then no longer be calculated by $p_n - p_w$. Both experiments and pore network models generating $p_n - p_w$–saturation relationships have been performed under dynamic conditions (see Bottero et al. [2006], Mirzaei and

Das [2007], DiCarlo [2004]), i.e. $p_n - p_w$-saturation relationships have been obtained as described above with the only difference that the system was not given time to reach equilibrium. Then, interestingly, $p_n - p_w$ values outside the static bounding capillary pressure–saturation curves are measured.

Fig. 2.14 shows the situation in more detail. Depending on the time rate of change of saturation, different "dynamic" $p_n - p_w$-saturation relationships will be obtained. Let us start considering a drainage process. Then, using the first of the above-described measurement techniques, a sudden increase in $p_n - p_w$ will not lead to an immediate decrease in S_w. Instead, after increasing $p_n - p_w$, S_w will be higher than the equilibrium saturation value which lies on the static $p_c(S_w)$ curve. This can be seen by comparison of curve 1 with the static drainage curves. If the drainage process is continued dynamically, i.e. without allowing the system to equilibrate, the dynamic drainage curve will lie above the static drainage curve for all values of $p_n - p_w$ (see curve 1 in Fig. 2.14). Depending on the time rate of change of saturation (or depending on the size of the sudden stepwise increase of $p_n - p_w$) the dynamic effect will be higher or lower. Curve 2 indicates larger dynamic effects than curve 1, implying a higher time rate of change of saturation or a larger stepwise increase of $p_n - p_w$. Similar considerations and experiments can be made for imbibition. Here, the dynamic curve lies below the static imbibition curves. Curve I is characterized by smaller dynamic effects than curve II.

Measuring $p_n - p_w$-saturation curves under various different dynamic conditions leads to the situation that is sketched in the right-hand picture of Fig. 2.14: the set of possible $p_n - p_w$-saturation data points is not bounded by main (or primary) drainage and main imbibition curve. Instead, possible $p_n - p_w$-saturation states can be found in the whole area $[S_{wr}; 1.0] \times [0; \infty]$. This makes the classical concept of capillary pressure–saturation functions appear even more dubious.

Several researchers have already put effort into the determination and quantification of dynamic effects in the $p_n - p_w$-saturation curve. The first person to determine and describe this effect was Stauffer [1978]. He experimentally investigated the influence of dynamic capillary pressure. Later, Hassanizadeh and Gray [1990, 1993b] studied two-phase flow in a theoretical framework based on rational thermodynamics. They could show that the classically proposed relationship, $p_c = p_n - p_w$ on the macro-scale, is only valid under static conditions. Under dynamic conditions, there is a

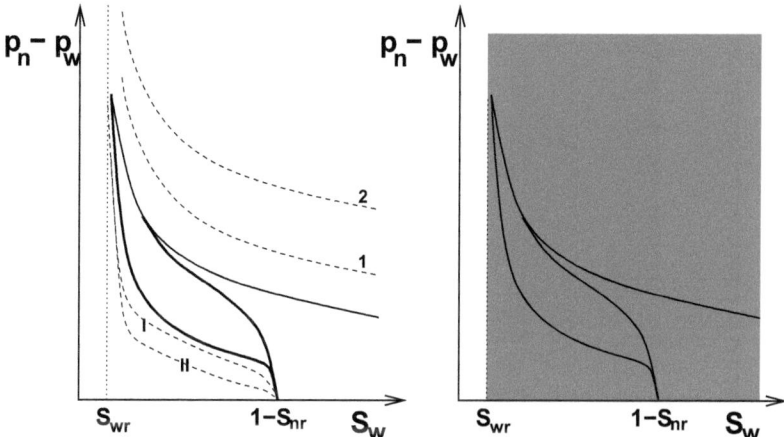

Figure 2.14: Impact of dynamic effects on $p_n - p_w$-saturation relationships. Left hand side: dynamic $p_c(S_w)$ curves (dashed lines) lie outside of the "bounding" main drainage and main imbibition curves; right hand side: region of possible $p_c(S_w)$ data points.

difference which is proportional to the time rate of change of saturation,

$$p_c = p_n - p_w + \tau \frac{\partial S_w}{\partial t}. \tag{2.88}$$

Here, τ might be a constant or a function of saturation and potentially also other parameters. Eq. (2.88) shows that under no flow conditions where $\frac{\partial S_w}{\partial t} = 0$, the two formulations are equivalent. An increasing number of experimentalists and modelers have been trying to quantify τ (see e.g. Bottero et al. [2006], Manthey et al. [2005], Mirzaei and Das [2007], DiCarlo [2004, 2005]).

An extremely crucial issue whose impact is often underestimated is the definition of macro-scale capillary pressure. In Sec. 2.2.5, it has been discussed that capillary pressure is originally a micro-scale quantity, namely the pressure jump across an interface between wetting and non-wetting phase. On the macro-scale, the experimental determination of capillary pressure suggests that capillary pressure is equal to the difference in non-wetting and wetting phase pressures applied at the boundaries of a measurement cell. In this definition, macro-scale capillary pressure is not related to phase-boundaries any more, but to bulk phase pressures that are applied outside the volume of interest. Recent work has concentrated on how to define macro-scale capillary pressure in a physically based way. Nordbotten et al. [2008] came up with a new averaging type for capillary pressure, the centroid-corrected phase average. Korteland et al. [2009] compared different types of definitions of macro-scale capillary pressure including *Nordbotten's* centroid-corrected phase average. It turned out that capillary pressure–saturation curves are extremely sensitive with respect to the definition of macro-scale capillary pressure.

In conclusion, a number of essential questions have arisen in connection to classical capillary pressure–saturation relationships:

- How can macro-scale capillary pressure be defined in a physically meaningful way?
- The capillary pressure–saturation relationship is hysteretic. Is the assumption reasonable that $p_n - p_w$ is a function of saturation only? What are other dependencies?
- Is a static capillary pressure–saturation relationship useful under dynamic flow conditions? How can dynamics be handled in a physically reasonable way?

2.5 Summary

In this chapter, the classical macro-scale model for two-phase flow in porous media has been presented and discussed. Therefore, scales and relevant parameters have been defined, fundamental balance equations have been formulated on the macro-scale (mass, momentum, and energy), and constitutive relationships and equations of state have been introduced. By showing the origin of equations (e.g. of Darcy's law) and of constitutive equations it was made obvious that a number of functional relationships in the classical two-phase flow model have been obtained empirically. It has been shown that this procedure leads to a number of shortcomings of the classical two-phase flow model. In that spirit, it was discussed that the application of the classical model is limited to local equilibrium situations, both in the sense of chemical equilibrium and thermal equilibrium, unless empirical relationships are employed. From these considerations, a strong need for a physically based and consistent model for flow and transport in porous media can be deduced. Such a model will be presented in the next chapter.

3 Alternative approach—an interfacial-area-based model

Based on thermodynamic considerations, Hassanizadeh and Gray [1980, 1990, 1993a,b] came up with an alternative model for two-phase flow in porous media. Their approach was based on averaging not only balance equations for bulk phases, but also balance equations for phase interfaces and common lines from the pore scale to the macro scale and on introducing the entropy inequality on the macro scale. In this chapter, their approach will be presented in detail as it represents the basis for macro-scale modeling using the interfacial area-based model. Therefore, in Sec. 3.1, the problematic issues arising in the classical two-phase flow model that have been discussed in Chap. 2 will be summarized and an overview of alternative approaches for two-phase flow modeling in porous media will be given. The remaining part of this work focusses on one of the alternative approaches, namely the above-mentioned rational thermodynamics approach of *Hassanizadeh and Gray*. Next, in Sec. 3.2, balance equations based on this approach will be formulated on the pore scale and upscaled to the macro scale. Constitutive relationships are derived on the macro scale (Sec. 3.3) to close the system of equations. In Sec. 3.4 and 3.5, the rational thermodynamics approach will be extended to specific physical situations: kinetic interphase mass transfer (Sec. 3.4) and kinetic interphase mass and energy transfer (Sec. 3.5). Finally, a chapter summary is given in Sec. 3.6.

3.1 Introduction

In this section, problematic issues of the current two-phase flow theory that have arisen during the introduction of the classical two-phase flow approach in Chap. 2 will be summarized. In order to remedy these shortcomings, a number of alternative approaches has been proposed. The summary of the problematic issues of the classical two-phase flow approach is given in Sec. 3.1.1 and an overview of alternative approaches to the classical two-phase flow model is given in Sec. 3.1.2.

3.1.1 Overview of problematic issues of the classical two-phase flow approach

The classical approach to model two-phase flow in porous media as outlined in Chap. 2 is to formulate mass balance equations for both fluid phases, simplify the two momentum balances to Darcy's Law, and—in case of non-isothermal systems—formulate one effective energy balance equation for the fluid–solid mixture. In order to close the system of equations, equilibrium laws are employed to account for the composition of the phases and it is postulated, that capillary pressure is a function of saturation only. This simple model has been obtained by empirical extensions of simple empirical models and subsequently been applied to more and more complex systems. Sometimes, however, these extensions did not have a sound physical basis (see e.g. Rose [2000]). Instead, existing equations were generalized by adding factors and making parameters and constitutive relationships phase- and / or component-dependent.

This procedure has proven useful and sufficient for a large number of engineering purposes and a large number of measurement techniques is available to determine the constitutive relationships of that model. However, there is a number of problematic issues related to the classical model which can be considered as common knowledge. Gray and Hassanizadeh [1991] could even show that the oversimplified concepts and theories of the classical approach do not only lack a sound physical basis, but even lead to a number of paradoxes and inconsistencies. In the following, we will discuss the problematic issues of the classical model related to:

1. Darcy's Law

2. Macro-scale capillary pressure, and

3. Interphase mass and energy transfer.

1. Darcy's Law
The original Darcy's Law was proposed as an empirical relationship linearly relating the specific flux through a column to the water head difference at the two sides

of the column assuming a homogeneous and isotropic porous medium and one-dimensional isothermal flow,

$$v = -\frac{1}{\mu}K\left(\frac{\Delta p}{\Delta x} + \rho g\right), \qquad (3.1)$$

where μ [Pa·s] is the dynamic viscosity of water, Δp [Pa] is the pressure difference, Δx [m] is the length of the column, K [m^2] is the intrinsic permeability, ρ [$\frac{kg}{m^3}$] is density, g [$\frac{m}{s^2}$] is gravity, and v [$\frac{m}{s}$] is the Darcy velocity, see Sec. 2.3.2.2. The same form of Darcy's Law is used nowadays to calculate flow velocities of fluid phases in physically highly complex systems, i.e. for non-isothermal compositional, multi-dimensional multi-phase flow processes in heterogeneous and anisotropic porous media. The only changes are that parameters are made phase and / or component dependent and factors are added which leads to

$$\underline{v}_\alpha = -\underline{\underline{K}} \frac{k_{r\alpha}(S_\alpha, \ldots)}{\mu_\alpha(p_\alpha, T, C_\alpha^\kappa)} \left(\nabla p_\alpha - \rho_\alpha(p_\alpha, T, C_\alpha^\kappa)\underline{g}\right), \qquad (3.2)$$

where α denotes a fluid phase, T [K] is temperature, C_α^κ [$\frac{kg}{m^3}$] is the concentration of component κ in phase α, S_α [−] is the saturation of phase α, and $k_{r\alpha}$ [−] is relative permeability. Eq. (3.2) indicates that in this physically highly complex situation, still, the only driving forces for flux of a phase are pressure difference and gravity. It seems that the effect of other relevant driving forces has to be absorbed by the relative permeability (see discussion below) such that the linearity between flux and pressure gradient does actually hold.

As given in Eq. (3.2), relative permeability is introduced as a scaling parameter, reducing the permeability of one phase due to the presence of another phase. However, this parameter does not come from any balance equations or an averaging process; instead, it is an empirical function. It is well known that this function $k_{r\alpha}(S_\alpha)$ is also hysteretic. But even worse: it has been shown, both experimentally and numerically, that $k_{r\alpha}$ is a function of a list of other parameters, including flow velocity, viscosity ratios, and boundary conditions (Miller et al. [1998], Demond and Roberts [1987], Avraam and Payatakes [1995a,b], Lefebvre du Prey [1973], Jerauld and Salter [1990])! Although relative permeability values should be within the range $[0\ldots1]$, relative permeability values > 1 have been reported from various experiments, see Berg et al. [2008] and a long list of references therein.

2. Macro-scale capillary pressure

Another critical point of classical two-phase flow models is the treatment of capillary pressure. While pore-scale capillary pressure is defined as the pressure drop across fluid–fluid interfaces at equilibrium and is directly related to physical interface properties, such as the curvature of the interface and interfacial tension, the meaning and the relation of macro-scale capillary pressure to other variables is not clear a priori. In classical models, macro-scale capillary pressure is usually defined as the difference between averaged phase pressures, see Sec. 2.2.5. The averaging region or volume is related to the size of a representative elementary volume (REV). Thus, this quantity which is from the physical point of view an interface-related parameter, is now defined with respect to a volume. The classical model of two-phase flow is given by

$$\frac{\partial (\phi \rho_\alpha S_\alpha)}{\partial t} - \nabla \cdot \left[\rho_\alpha \underline{\underline{K}} \frac{k_{r\alpha}(S_\alpha)}{\mu_\alpha} \left(\nabla p_\alpha - \rho_\alpha g \right) \right] = F_\alpha, \qquad (3.3)$$

where an extended form of Darcy's Law given in Eq. (3.2) has been inserted into the mass balance equation for each phase α. Here, ϕ [–] is porosity and F_α [$\frac{kg}{m^3 \cdot s}$] is an external source or sink of phase α (e.g. injection or pumping). As Eq. (3.3) represents two equations for the four unknowns S_α and p_α of both phases, two additional closure relationships are needed. One of them is obviously $S_w + S_n = 1$, where w denotes the wetting phase and n denotes the non-wetting phase. But the second one needs to be found. As capillary pressure and saturation are known to be correlated, the classical model is closed by postulating

$$p_c := p_n - p_w := p_c(S_w). \qquad (3.4)$$

However, Eq. (3.4) is problematic, as the relation $p_c(S_w)$ is highly hysteretic. In numerous experimental studies measuring non-wetting and wetting phase pressure along with saturation under no-flow conditions (static experiments), it has been shown that—if capillary pressure is defined as in Eq. (3.4)—it is dependent on the process (i.e. whether a porous medium is being drained or imbibed) and on the saturation history. This typically leads to a situation as shown in Fig. 3.1: during different drainage and imbibition cycles, all capillary pressure–saturation values between a bounding drainage and a bounding imbibition function are possible! Fig. 3.1 brings up the following question: is saturation really the only parameter on which capillary pressure depends?

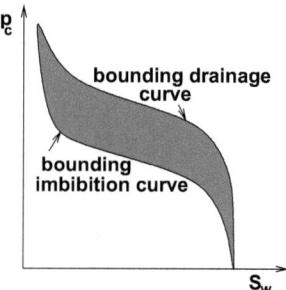

Figure 3.1: Hysteresis of capillary pressure: all (S_w, p_c) values between main imbibition and main drainage curve are physically possible.

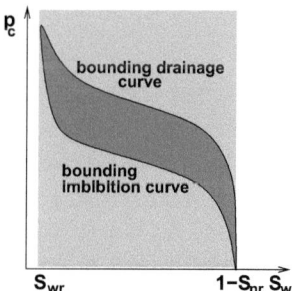

Figure 3.2: In the dynamic case, (S_w, p_c) values even outside the static main imbibition and main drainage curve are physically possible.

The situation becomes even worse if one accounts for the fact that most often, dynamic situations are of interest where non-zero flow velocities occur. In that case, experimental studies show that capillary pressure values *outside* of the bounding drainage and bounding imbibition curve determined under static conditions occur! This means that in principle, all capillary pressure–saturation values (in between the residual saturations) in the positive capillary pressure–saturation plane represent physically possible states, see Fig. 3.2! This clearly indicates that capillary pressure be a function of not only saturation and that the classical approach is at best incomplete!

3. Interphase mass and energy transfer

Interphase mass transfer takes place across the fluid–fluid interface. Therefore, in the case of kinetic mass transfer, it is highly dependent on fluid–fluid interfacial

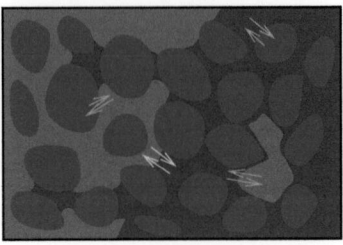
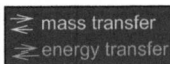

Figure 3.3: Interphase mass and energy transfer.

area. Although this fact is widely acknowledged, classical models do not and cannot account for this fact as interfacial area is unknown. That means that it is not possible to account for interphase mass transfer for kinetic processes in a physically-based manner. Instead, classical models either describe interphase mass and energy transfer by equilibrium relations or they use empirical models to describe the kinetics. Similar considerations can be made with respect to interphase energy transfer. But here, additionally to fluid–fluid interfaces, solid–fluid interfaces need to be taken into account as energy can also be transferred from and to the solid phase, see Fig. 3.3. This is not possible for mass unless sorption is considered which is not the case in this work. Can these problems be overcome and can interphase mass and energy transfer be modeled based on physically-motivated equations?

3.1.2 Overview of alternative approaches

The above questions have led to research activities questioning the physical correctness of classical two-phase flow models. Alternative conceptual and mathematical models have been developed which describe two-phase flow based on thermodynamic principles. Among these are an approach based on mixture theory (Bowen [1982]), a rational thermodynamics approach by Hassanizadeh and Gray [1980, 1990, 1993a,b], a thermodynamically constrained averaging theory (Gray and Miller [2005]), and an approach based on averaging and non-equilibrium thermodynamics by Marle [1981] and Kalaydjian [1987]. Another alternative model separating percolating and non-percolating part of each phase was suggested by Hilfer [2006]. These models are briefly sketched in the following:

1. **mixture theory approach by Bowen [1982]**. The approach by Bowen [1982] is based on mixture theory which implies that balance equations are formulated directly on the macro scale. The entropy inequality (second law of thermodynamics) is introduced on the macro scale in order to derive constitutive relationships. Bowen [1982] shows how the classical two-phase flow model can be obtained if certain assumptions are imposed.

2. **rational thermodynamics approach by Hassanizadeh and Gray [1980, 1990, 1993a,b]**. The rational thermodynamics approach proceeds from the formulation of pore-scale balance equations (for mass, momentum, energy, and entropy) for not only the bulk fluid phases, but also for phase-interfaces and common lines. Then, the system of balance equations is averaged to the macro scale. The entropy inequality is introduced on the macro scale and used to derive constitutive equations. An important characteristic of this alternative theory is that capillary pressure is not a function of saturation only, but also depends on interfacial areas. Also, the difference in macro-scale phase pressures is not equal to capillary pressure if saturations are changing; instead the difference between phase pressures deviates from capillary pressure by a dynamic term.

3. **thermodynamically constrained averaging theory by Gray and Miller [2005]**. The approach based on thermodynamically constrained averaging is similar in its principle ideas to the rational thermodynamics approach of *Hassanizadeh and Gray*. One difference lies—as the name indicates—in the thermodynamic formulation (thermodynamically constrained averaging instead of rational thermodynamics). The main difference is the scale where the entropy inequality and constitutive relationships are introduced. While in the rational thermodynamics approach they are introduced on the macro scale, in the thermodynamically contrained averaging approach, they are introduced on the pore-scale and averaged to the macro scale.

4. **averaging and non-equilibrium thermodynamics by Marle [1981] and Kalaydjian [1987].** The approach of *Marle and Kalaydjian* is based on the formulation of balance equations for mass and momentum on the macroscopic scale (they consider isothermal systems so no energy balance equations are needed). The entropy inequality is introduced on the macro scale in the frame of irreversible thermodynamics. In order to close their

system of equations, they introduce a number of phenomenological equations.

5. **approach based on formulation of balance equations for percolating and non-percolating fraction of each phase by Hilfer [2006]**. *Hilfer* reasoned that it is impossible to rigorously upscale pore-scale balance equations to the macro scale and that at some point, all of the upscaling approaches need to rely on assumptions. Therefore, he argues that it is from the physical point of view equally promising or valid to directly postulate equations on the macro scale. Hilfer splits each fluid phase into percolating and non-percolating fraction and thus, formulates balance equations for four entities. The exchange terms between percolating and non-percolating fractions are modeled based on empirical relationships. An especially interesting feature of Hilfer's approach is that there is no necessity for a constitutive capillary pressure–saturation relationship as an input to the model; instead, the capillary pressure–saturation relationship is an outcome of the model.

All of these approaches try to describe two-phase flow differently from the classical approaches with the goal to resolve the above three problematic issues (or parts of them). Arguments can be found in favor and against all of these approaches. The goal of this work is a physically based macro-scale modeling of two-phase flow in porous media accounting for the important role of phase-interfaces. Without properly accounting for phase-interfaces, it is not only impossible to account for chemical and thermal non-equilibrium processes in a physically based way, but one also has to deal with hysteresis in constitutive relationships like capillary pressure. Therefore, a thermodynamically consistent approach is envisaged that also acknowledges the important role of interfacial area. While other choices may be equally justified, the rational therodynamics approach of *Hassanizadeh and Gray* will be used for the further studies.

3.2 Balance equations for two-phase flow

In this section, the basic principles and ideas of the rational thermodynamics approach of Hassanizadeh and Gray will be discussed for the case of isothermal immiscible two-phase flow in a porous medium (Sec. 3.2.1). Then, it will be shown how they formulated balance equations of mass, momentum, energy, and entropy on the pore scale and averaged them to the macro scale (Sec. 3.2.2 through 3.2.5).

3.2.1 Basic ideas

One of the main ideas of Hassanizadeh and Gray's approach is to consider a porous medium occupied by two different fluid phases as a set of different continua or entities which exchange mass, momentum, energy, and entropy. The novelty of their approach was to not only consider bulk fluid phases as entities, but to acknowledge the fact that phase-interfaces and common lines may behave in a completely different way. Therefore, in a porous medium occupied by two fluid phases, seven such entities can be identified: three bulk phases (wetting, non-wetting, and solid phase), three phase-interphases (wetting–non-wetting, wetting–solid, and non-wetting–solid interface), and one common line where all three phases (and all three phase-interfaces) meet.

This situation is illustrated in a simplified 2d version in Fig. 3.4 where two fluid phases in a capillary tube are shown (might be a pore throat). Of course, in a 3d porous medium, these bulk phases would represent volumetric entities. The interfaces are located in between the bulk phases: the ws-interface separates wetting fluid phase and solid phase, the ns-interface is between non-wetting fluid phase and solid phase, and the wn-interface separates the two fluid phases. Considering these interfaces in 3d, they would actually represent three-dimensional entities, because they are actually a few molecules thick. However, their thickness is very small compared to the other two dimensions. Whether these interfaces can be considered as lower-dimensional objects will be discussed later in this section. Finally, the set of points where all three phases and all three interfaces meet is called the wns-common line. In the figure, it reduces to a point; in a 3d porous medium, it is strictly speaking a three-dimensional object. However, its extension in the two lateral space direction is very small compared to its extension in longitudinal direction. In the capillary tube shown in Fig. 3.4, for example, the common line would form a circle on the cylinder surface.

An important issue in this respect is that often, the wetting fluid forms a thin layer ("film") on the solid surface, see Fig. 3.5. In that sense, an ns-interface would in fact be a combination of a ws- and a wn-interface. For many two-phase flow situations in porous media, a wetting-phase film can be interpreted as a layer which modifies the properties of the ns-interface; alternatively, for many applications, its influence can be completely neglected.

Hassanizadeh and Gray's idea was to formulate balance equations for mass, mo-

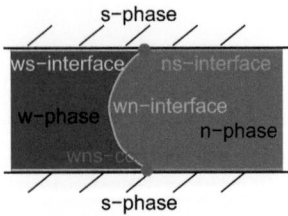

Figure 3.4: Definition of phases, interfaces, and common lines.

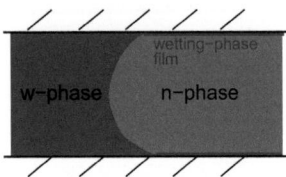

Figure 3.5: Wetting fluid film on the solid surface.

mentum, energy, and entropy for all seven entities on the pore-scale, volume-average them over representative elementary volumes to the macroscopic scale, and introduce a rational thermodynamics approach and the entropy inequality on the macro scale to formulate constitutive equations. In order to formulate the pore-scale balance equations it is essential to precisely define the terms phase, interface, and common line. Therefore, Fig. 3.4 is revisited, but this time considered on the molecular scale where different particles can be identified, see Fig. 3.6. Let us consider exemplarily, how a wetting-non-wetting interface is defined: Fig. 3.6 shows that there are regions where particles only "see" particles of their own kind. These regions are called the non-wetting phase and the wetting-phase respectively. Then, there is a region where particles see particles of both kinds. This is the region which

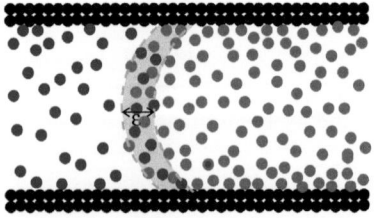

Figure 3.6: Two fluids in a capillary tube—molecular-scale picture.

is called wn-interface and which clearly has other physical properties than bulk w- and n-phase. Similar considerations can be made for the ws- and ns-interfaces as well as for the wns-common line. In case solubility effects play a role, a small percentage of particles of the other kind needs to be allowed in the respective phase.

It remains to identify the dimensionality of the entities. Clearly, bulk phases are volumetric objects. Interfaces have a very small thickness ε as they are only a few layers of particles thin compared to the other dimensions, e.g. the diameter of the capillary tube. As an illustration, the thickness of a water–vapor interface is 2.5-2.8 angstrom (Schwartz et al. [1990]) while the dimensions of a pore throat are typically in the range of tenths of millimeters. This comparison shows that interfaces may indeed be considered as two-dimensional objects in good approximation. Similar considerations can be made for the common lines to argue that they represent one-dimensional objects. Having these issues in mind it becomes obvious that it is indeed possible to formulate balance equations for all seven entities and still consider interfaces as two-dimensional and common lines as one-dimensional objects: the particles that make up an interface have a certain mass, momentum, energy, and entropy which can be balanced, but still the thickness of the interface can be considered negligible. These considerations lead to areal balance equations for interfaces and analogous considerations lead to 1d balance equations for common lines.

In the following, balance equations for phases and interfaces will be focussed on. Balance equations for common lines could be obtained analogously. However, due to their small mass, momentum, and energy compared to mass, momentum, and energy of bulk phases and interfaces, they are neglected here. Also, their consideration and discussion would additionally blow up the set of equations. In the remaining part of this section, the principle averaging strategy for general balance equations will be outlined for both phases and interfaces.

3.2.1.1 Volume averaging of general phase balances

The fundamentals for the rational thermodynamics approach of Hassanizadeh and Gray were laid in the late 1970's. Hassanizadeh and Gray [1979a] developed an averaging procedure to average phase balance equations for two-phase flow from the pore scale to the macro scale in a consistent way. Therefore, they started off with a general conservation equation for a property ψ on the pore scale,

$$\frac{\partial (\rho\psi)}{\partial t} + \nabla \cdot (\rho\underline{v}\psi) - \nabla \cdot \underline{i} - \rho f = \rho G, \qquad (3.5)$$

where ρ is a mass density function, $\underline{v}$ is velocity, and the properties $\underline{i}$, f, and G occur in the different balance laws and have a special meaning in each of the laws: $\underline{i}$ is a general diffusive term, f represents general body forces, and G is a source / sink term. Eq. (3.5) is supplemented by a continuity equation for interfaces. At an interface $\alpha\beta$, fluxes must be continuous if mass does not accumulate on the interface,

$$\left[\rho_\alpha \psi_\alpha \left(\underline{v}_{\alpha\beta} - \underline{v}_\alpha\right) + \underline{i}_\alpha\right] \cdot \underline{n}_{\alpha\beta} + \left[\rho_\beta \psi_\beta \left(\underline{v}_{\alpha\beta} - \underline{v}_\beta\right) + \underline{i}_\beta\right] \cdot \underline{n}_{\beta\alpha} = 0, \qquad (3.6)$$

where $\underline{v}_{\alpha\beta}$ is the velocity of the interface and $\underline{n}_{\alpha\beta}$ is the unit normal vector pointing out of the α-phase into the β-phase and $\underline{n}_{\beta\alpha} = -\underline{n}_{\alpha\beta}$. The lower indices α and β indicate on which side of the interface the respective terms are evaluated.

Averaging Eq. (3.5) subject to Eq. (3.6) over a representative elementary volume, Hassanizadeh and Gray [1979a] obtained the following macro-scale balance equation,

$$\frac{\partial}{\partial t}\left(\langle\rho\rangle_\alpha \bar{\psi}_\alpha\right) + \nabla \cdot \left(\langle\rho\rangle_\alpha \bar{\underline{v}}_{\alpha,s} \bar{\psi}_\alpha\right) - \nabla \cdot \bar{\underline{i}}_\alpha - \langle\rho\rangle_\alpha \bar{f}_\alpha - \langle\rho\rangle_\alpha e_\alpha (\rho\psi)$$
$$- \langle\rho\rangle_\alpha \hat{I}_\alpha = \langle\rho\rangle_\alpha \bar{G}_\alpha \qquad (3.7)$$

subject to the constraint that the overall mass transfer must be zero,

$$\sum_\alpha \langle\rho\rangle_\alpha \left[e_\alpha (\rho\psi) + \hat{I}_\alpha\right] = 0, \qquad (3.8)$$

and

$$\left[\langle\rho\rangle_\alpha \bar{\psi}_\alpha \left(\bar{\underline{v}}_{\alpha,s} - \bar{\underline{v}}_{\alpha\beta,s}\right) - \underline{i}_\alpha\right] \cdot \underline{n}_{\alpha\beta} = 0 \quad \text{for phase interfaces within an REV.} \quad (3.9)$$

Here, the brackets $\langle\cdot\rangle$ denote a volume averaging operator, the overbars denote a mass average operator, $e_\alpha (\rho\psi)$ is the density function due to phase change defined by

$$e_\alpha (\rho\psi) = \frac{1}{\langle\rho\rangle_\alpha \, dV} \sum_{\beta\neq\alpha} \int_{dA_{\alpha\beta}} \rho\psi \left(\underline{v}_{\alpha\beta} - \underline{v}_\alpha\right) \cdot \underline{n}_{\alpha\beta} dA, \qquad (3.10)$$

and $\hat{I}$ is the density function for exchange of ψ between phase α and β through mechanical interactions. The vector $\underline{n}_{\alpha\beta}$ denotes the unit normal vector of the macro-scale $\alpha\beta$-interface, $\bar{\underline{v}}_{\alpha,s}$ is the macro-scale seepage velocity of the α-phase, and $\bar{\underline{v}}_{\alpha\beta,s}$ is the macro-scale velocity of the $\alpha\beta$-interface. Note that $\langle \rho \rangle_\alpha$ is related to well-known macro-scale properties via

$$\langle \rho \rangle_\alpha = \phi \rho_\alpha S_\alpha . \tag{3.11}$$

3.2.1.2 Volume averaging of general interface balances

Gray and Hassanizadeh [1989] studied averaging theorems for phase-interfaces. Applying them to the general pore-scale balance equation for interfaces, they upscaled these balance equations to the macro scale. Similarly to the work of Hassanizadeh and Gray [1979a], they started off with a general balance equation for interfaces at the pore scale,

$$\frac{\partial (\Gamma \psi_{\alpha\beta})}{\partial t} + \nabla \cdot \left(\Gamma \psi_{wn} \underline{v}_{\alpha\beta} \right) - 2\Gamma \underline{v}_{\alpha\beta} \cdot \underline{n}_{\alpha\beta} K_M \psi_{wn} - \nabla \cdot \underline{i}_{\alpha\beta} - \Gamma f_{\alpha\beta} - \Gamma G_{\alpha\beta}$$
$$= -\sum_{\alpha=1}^{2} \left[\rho_\alpha \psi_\alpha \left(\underline{v}_{\alpha\beta} - \underline{v}_\alpha \right) + \underline{i}_\alpha \right] \cdot \underline{n}_\alpha, \tag{3.12}$$

where Γ is the areal mass density, K_M is the mean curvature of the interface, and ψ, $\underline{i}$, f, and G are defined as above, but as properties of interfaces instead of phases. This implies, that $i_{\alpha\beta}$ is the flux within the surface, $f_{\alpha\beta}$ is the external supply term for the surface, and that $G_{\alpha\beta}$ is the production term within the surface. In their paper, Gray and Hassanizadeh [1989] develop necessary tools and a systematic averaging procedure to average the pore-scale interface equation, Eq. (3.12), over the union of interfaces within a representative elementary volume to come up with a macro-scale interface balance equation. In case common lines are not considered—in their work, Gray and Hassanizadeh [1989] derived their balance equation also for the case where common lines are considered—the following macro-scale balance equation for interfaces is obtained,

$$\frac{\partial}{\partial t}\left(\Gamma_{\alpha\beta}\bar{\psi}_{\alpha\beta}a_{\alpha\beta}\right) + \nabla \cdot \left(\Gamma_{\alpha\beta}\bar{\underline{v}}_{\alpha\beta,s}\bar{\psi}_{\alpha\beta}a_{\alpha\beta}\right) - \nabla \cdot \left(\bar{i}_{\alpha\beta}a_{\alpha\beta}\right)$$
$$- \Gamma_{\alpha\beta}\bar{f}_{\alpha\beta}a_{\alpha\beta} - \Gamma_{\alpha\beta}\bar{G}_{\alpha\beta}a_{\alpha\beta}$$
$$= \sum_{\alpha=1}^{2} \frac{1}{A_{\alpha\beta}} \int_{A_{\alpha\beta}} \left[\rho_{\alpha}\psi_{\alpha}\left(\underline{v}_{\alpha} - \underline{v}_{\alpha\beta}\right) - \underline{i}_{\alpha}\right] \cdot \underline{n}_{\alpha\beta}\, dA,$$

where $A_{\alpha\beta}$ denotes the interface separating phase α from phase β and $a_{\alpha\beta}$ is the specific $\alpha\beta$-interfacial area (the sum of the areas of all $\alpha\beta$-interfaces per volume of REV). Note that the expression $\Gamma_{\alpha\beta}a_{\alpha\beta}$ was obtained from

$$\langle\Gamma\rangle_{\alpha\beta} = \Gamma_{\alpha\beta}a_{\alpha\beta} \,. \tag{3.13}$$

In the following sections (Sec. 3.2.2 through 3.2.5), the general macro-scale balance equations will be applied and specifically discussed for mass, momentum, energy, and entropy according to Hassanizadeh and Gray [1979b].

3.2.2 Mass balance

In order to obtain the macro-scale mass balance equation, Hassanizadeh and Gray [1979b] replaced the variables of the general mass balance equation for **phases** by

$$\psi = 1, \qquad \underline{i} = \underline{0}, \qquad f = 0, \qquad G = 0. \tag{3.14}$$

Thus, they obtained the following macro-scale mass balance equation for phases,

$$\frac{\partial \langle\rho\rangle_{\alpha}}{\partial t} + \nabla \cdot \left(\langle\rho\rangle_{\alpha}\bar{\underline{v}}_{\alpha,s}\right) = \langle\rho\rangle_{\alpha}e_{\alpha}\left(\rho\right) \tag{3.15}$$

subject to

$$\sum_{\alpha} \langle\rho\rangle_{\alpha}e_{\alpha}\left(\rho\right) = 0 \tag{3.16}$$

and

$$\left[\langle\rho\rangle_{\alpha}\left(\bar{\underline{v}}_{\alpha,s} - \bar{\underline{v}}_{\alpha\beta,s}\right)\right] \cdot \underline{n}_{\alpha\beta} = 0 \qquad \text{for phase interfaces within an REV.} \tag{3.17}$$

Similarly to the procedure of Hassanizadeh and Gray [1979b], Gray and Hassanizadeh [1989] specified the general mass balance equation for **interfaces** by setting

$$\psi_{\alpha\beta} = 1, \quad \underline{i}_{\alpha\beta} = \underline{0}, \quad f_{\alpha\beta} = 0, \quad G_{\alpha\beta} = 0, \quad \psi_\alpha = 1, \quad \underline{i}_\alpha = \underline{0}. \quad (3.18)$$

Using these definitions, they ended up with the macro-scale mass balance equation for an $\alpha\beta$-interface,

$$\frac{\partial}{\partial t}\left(\Gamma_{\alpha\beta} a_{\alpha\beta}\right) + \nabla \cdot \left(\Gamma_{\alpha\beta} \bar{\underline{v}}_{\alpha\beta,s} a_{\alpha\beta}\right)$$

$$= \sum_{\alpha=1}^{2} \frac{1}{A_{\alpha\beta}} \int_{A_{\alpha\beta}} \left[\rho_\alpha \left(\underline{v}_\alpha - \underline{v}_{\alpha\beta}\right)\right] \cdot \underline{n}_{\alpha\beta} \, dA. \quad (3.19)$$

3.2.3 Momentum balance

Following the work of Hassanizadeh and Gray [1979b], the following substitutions need to be made in order to obtain the general momentum balance equation for **phases**, where ψ, f, and G need to be considered vector-type properties now and $\underline{i}$ a matrix-type quantity:

$$\underline{\psi} = \underline{v}, \quad \underline{\underline{i}} = \underline{\underline{t}}, \quad \underline{f} = \underline{g}, \quad \underline{G} = \underline{0}, \quad (3.20)$$

where $\underline{\underline{t}}$ is the stress tensor and $\underline{g}$ is the external supply of momentum. Inserting these substitutions into the general momentum balance equation, they obtained the following macro-scale momentum balance equation for phases,

$$\frac{\partial}{\partial t}\left(\langle\rho\rangle_\alpha \bar{\underline{v}}_{\alpha,s}\right) + \nabla \cdot \left(\langle\rho\rangle_\alpha \bar{\underline{v}}_{\alpha,s} \bar{\underline{v}}_{\alpha,s}^T\right) - \nabla \cdot \underline{\underline{t}}_\alpha - \langle\rho\rangle_\alpha \bar{\underline{g}}_\alpha - \langle\rho\rangle_\alpha \underline{e}_\alpha \left(\rho \tilde{\underline{v}}\right)$$

$$- \langle\rho\rangle_\alpha \underline{e}_\alpha \left(\rho\right) \bar{\underline{v}}_{\alpha,s} - \langle\rho\rangle_\alpha \hat{\underline{T}}_\alpha = 0 \quad (3.21)$$

subject to

$$\sum_\alpha \langle\rho\rangle_\alpha \left(\underline{e}_\alpha \left(\rho \tilde{\underline{v}}\right) - \underline{e}_\alpha \left(\rho\right) \bar{\underline{v}}_{\alpha,s} - \hat{\underline{T}}_\alpha\right) = 0 \quad (3.22)$$

and

$$\left[\langle\rho\rangle_\alpha \bar{\underline{v}}_\alpha \left(\bar{\underline{v}}_{\alpha,s} - \bar{\underline{v}}_{\alpha\beta,s}\right) - \underline{\underline{t}}_\alpha\right] \cdot \underline{n}_{\alpha\beta} = 0 \quad \text{for phase-interfaces within an REV.} \quad (3.23)$$

Here, velocity is split in two parts, an average and a fluctuating part, by $\underline{v} = \bar{\underline{v}} + \tilde{\underline{v}}$, $\hat{\underline{T}}_\alpha$ is the direct momentum transfer, and $\underline{e}_\alpha\,(\rho\tilde{\underline{v}})$ is the momentum exchange term due to mass transfer.

Gray and Hassanizadeh [1989] used their general averaged equation for interfaces to specify them to become the momentum balance equation for **interfaces** by interpreting the parameters $\psi_{\alpha\beta}$, $f_{\alpha\beta}$, $G_{\alpha\beta}$, as well as ψ_α as vector-type and $\underline{i}_{\alpha\beta}$ as well as $\underline{i}_\alpha$ as matrix-type properties and by setting

$$\underline{\psi}_{\alpha\beta} = \underline{v}_{\alpha\beta}, \quad \underline{\underline{i}}_{\alpha\beta} = \underline{\underline{t}}_{\alpha\beta}, \quad \underline{f}_{\alpha\beta} = \underline{g}_{\alpha\beta}, \quad \underline{G}_{\alpha\beta} = \underline{0}, \quad \underline{\psi}_\alpha = \underline{v}_\alpha, \quad \underline{\underline{i}}_\alpha = \underline{\underline{t}}_\alpha. \tag{3.24}$$

Using these definitions, they ended up with the macro-scale momentum balance equation for an $\alpha\beta$-interface,

$$\frac{\partial}{\partial t}\left(\Gamma_{\alpha\beta}\bar{\underline{v}}_{\alpha\beta}a_{\alpha\beta}\right) + \nabla \cdot \left(\Gamma_{\alpha\beta}\bar{\underline{v}}_{\alpha\beta,s}\bar{\underline{v}}^T_{\alpha\beta,s}a_{\alpha\beta}\right) - \nabla \cdot \left(\underline{\underline{t}}_{\alpha\beta}a_{\alpha\beta}\right) - \Gamma_{\alpha\beta}\bar{\underline{g}}_{\alpha\beta}a_{\alpha\beta}$$

$$= \sum_{\alpha=1}^{2}\frac{1}{A_{\alpha\beta}}\int_{A_{\alpha\beta}}\left[\rho_\alpha \underline{v}_\alpha\left(\underline{v}_\alpha - \underline{v}_{\alpha\beta}\right) - \underline{\underline{t}}_\alpha\right]\cdot\underline{n}_{\alpha\beta}\,dA. \tag{3.25}$$

3.2.4 Energy balance

The terms for the energy balance equation become even more complex. Hassanizadeh and Gray [1979b] came up with the following substitutions of the variables in their general balance equation for **phases**:

$$\psi = u + \frac{1}{2}v^2, \quad \underline{i} = \underline{\underline{t}}\cdot\underline{v} + \underline{q}, \quad f = \underline{g}\cdot\underline{v} + h, \quad G = 0, \tag{3.26}$$

where u is the specific internal energy, v is the absolute value of velocity $\underline{v}$, $\underline{q}$ is the partial macroscopic heat vector, and h is the total macroscopic heat supply from the external world.

These definitions lead to the following macro-scale energy balance equations,

$$\frac{\partial}{\partial t}\left(\langle\rho\rangle_\alpha u_\alpha + \frac{1}{2}\langle\rho\rangle_\alpha \bar{v}_{\alpha,s}^2\right) + \nabla \cdot \left[\left(\langle\rho\rangle_\alpha u_\alpha - \frac{1}{2}\langle\rho\rangle_\alpha \bar{v}_{\alpha,s}^2\right)\bar{\underline{v}}_{\alpha,s}\right]$$
$$- \nabla \cdot \left(\underline{\underline{t}}_\alpha \cdot \bar{\underline{v}}_{\alpha,s} + \underline{q}_\alpha\right) - \langle\rho\rangle_\alpha \underline{\bar{g}}_\alpha \cdot \bar{\underline{v}}_{\alpha,s}$$
$$- \langle\rho\rangle_\alpha h_\alpha - \langle\rho\rangle_\alpha \hat{Q}_\alpha - \langle\rho\rangle_\alpha \hat{\underline{T}}_\alpha \cdot \bar{\underline{v}}_{\alpha,s} - \langle\rho\rangle_\alpha e_\alpha(\rho\hat{u})$$
$$- \langle\rho\rangle_\alpha e_\alpha(\rho\tilde{\underline{v}}) \cdot \bar{\underline{v}}_{\alpha,s} - \frac{1}{2}\langle\rho\rangle_\alpha e_\alpha(\rho)\bar{v}_{\alpha,s}^2 = 0 \qquad (3.27)$$

subject to

$$\sum_\alpha \langle\rho\rangle_\alpha \left(e_\alpha(\rho\hat{u}) + e_\alpha(\rho\tilde{\underline{v}}) \cdot \bar{\underline{v}}_{\alpha,s} + \frac{1}{2}\bar{v}_{\alpha,s}^2 e_\alpha(\rho) + \hat{\underline{T}}_\alpha \cdot \bar{\underline{v}}_{\alpha,s} + \hat{Q}_\alpha\right) = 0 \qquad (3.28)$$

and

$$\left[\langle\rho\rangle_\alpha \left(u_\alpha + \frac{1}{2}v_{\alpha,s}^2\right)(\bar{\underline{v}}_{\alpha,s} - \bar{\underline{v}}_{\alpha\beta,s}) - \underline{\underline{t}}_\alpha \cdot \bar{\underline{v}}_{\alpha,s} - \underline{q}_\alpha\right] \cdot \underline{n}_{\alpha\beta} = 0 \qquad (3.29)$$

for phase interfaces within an REV. Here, $e_\alpha(\rho\hat{u})$ and $\hat{Q}_\alpha$ account for exchange of energy through mass transfer and mechanical interactions, respectively.

For interfaces, Gray and Hassanizadeh [1989] came up with similar substitutions. They specified the variables in the general conservation equation for **interfaces** in the following way:

$$\begin{aligned}\psi_{\alpha\beta} &= u_{\alpha\beta} + \frac{1}{2}v_{\alpha\beta}^2 & \underline{i}_{\alpha\beta} &= \underline{\underline{t}}_{\alpha\beta} \cdot \underline{v}_{\alpha\beta} + \underline{q}_{\alpha\beta}, & f &= \underline{g}_{\alpha\beta} \cdot \underline{v}_{\alpha\beta} + h_{\alpha\beta}, & G_{\alpha\beta} &= 0 \\ \psi_\alpha &= u_\alpha + \frac{1}{2}v_\alpha^2, & \underline{i}_\alpha &= \underline{\underline{t}}_\alpha \cdot \underline{v}_\alpha + \underline{q}_\alpha.\end{aligned} \qquad (3.30)$$

Inserting these substitutions into their general conservation equation for interfaces, Gray and Hassanizadeh [1989] obtained the following macro-scale energy balance equation for interfaces,

$$\left[\frac{\partial}{\partial t}\left(\Gamma_{\alpha\beta}\left(\bar{\Pi}_{\alpha\beta}+\frac{1}{2}\bar{v}^2_{\alpha\beta,s}\right)a_{\alpha\beta}\right)+\nabla\cdot\left(\Gamma_{\alpha\beta}\left(\bar{\Pi}_{\alpha\beta}+\frac{1}{2}\bar{v}^2_{\alpha\beta,s}\right)\bar{v}_{\alpha\beta,s}a_{\alpha\beta}\right)\right.$$
$$-\nabla\cdot\left(\left(\underline{\underline{T}}_{\alpha\beta}:\bar{v}_{\alpha\beta,s}+\underline{Q}_{\alpha\beta}\right)a_{\alpha\beta}\right)$$
$$-\Gamma_{\alpha\beta}\left(\bar{g}_{\alpha\beta}\cdot\bar{v}_{\alpha\beta,s}+h_{\alpha\beta}\right)a_{\alpha\beta}$$
$$=-\sum_{\alpha=1}^{2}\frac{1}{A_{\alpha\beta}}\int_{A_{\alpha\beta}}\left[\rho_\alpha\left(u_\alpha+\frac{1}{2}v_\alpha^2\right)\right.$$
$$\left.\cdot(\underline{v}_\alpha-\underline{v}_{\alpha\beta})-\underline{\underline{t}}_\alpha\underline{v}_\alpha-\underline{q}_\alpha\right]\cdot\underline{n}_{\alpha\beta}\,dA,\quad(3.31)$$

with

$$\bar{\Pi}_{\alpha\beta}=\bar{u}_{\alpha\beta}+\frac{1}{2}\tilde{v}^2_{\alpha\beta} \tag{3.32}$$

$$\underline{Q}_{\alpha\beta}=\underline{Q}_{\alpha\beta}+\langle\underline{\underline{\tilde{t}}}_{\alpha\beta}\tilde{\underline{v}}_{\alpha\beta}\rangle_{\alpha\beta} \tag{3.33}$$

$$h_{\alpha\beta}=\bar{h}_{\alpha\beta}+\langle\tilde{\underline{v}}_{\alpha\beta}\cdot\tilde{\underline{g}}_{\alpha\beta}\rangle_{\alpha\beta} \tag{3.34}$$

The symbol ":" denotes the Frobenius inner product.

3.2.5 Entropy balance

Although it may not seem intuitive at first glance, it is possible to formulate a conservation equation for entropy. It is well-known that entropy can only increase or stay constant (second law of thermodynamics). This knowledge can be put into an entropy conservation equation when providing an entropy production term G which is restricted by $G \geq 0$.

In this spirit, Hassanizadeh and Gray [1979b] made the following substitutions in their general conservation equation for **phases** in order to come up with a conservation equation for entropy,

$$\psi=\eta,\qquad \underline{i}=\underline{\phi},\qquad f=b,\qquad G=\Lambda. \tag{3.35}$$

The variables have the following meanings: η is the specific internal entropy, $\underline{\phi}$ is the entropy flux vector, b is external supply of specific entropy, and Λ is the net production of entropy for which—according to the second law of thermodynamics—$\Lambda \geq 0$ holds.

Using these variables in the general conservation equation for phases, Hassanizadeh and Gray [1979b] obtained the following macro-scale entropy balance equation for phases,

$$\frac{\partial}{\partial t}\left(\langle \rho \rangle_\alpha \bar{\eta}_\alpha\right) + \nabla \cdot \left(\langle \rho \rangle_\alpha \bar{\eta}_\alpha \underline{\bar{v}}_{\alpha,s}\right) - \nabla \cdot \underline{\phi}_\alpha - \langle \rho \rangle_\alpha \left(\bar{b}_\alpha + e_\alpha\left(\rho \eta\right) + \hat{\phi}_\alpha\right) = \langle \rho \rangle_\alpha \bar{\Lambda}_\alpha \quad (3.36)$$

subject to

$$\sum_\alpha \langle \rho \rangle_\alpha \left(e_\alpha\left(\rho \eta\right) + \hat{\phi}_\alpha\right) = 0 \quad (3.37)$$

and

$$\left[\langle \rho \rangle_\alpha \bar{\eta}_\alpha \left(\underline{\bar{v}}_{\alpha,s} - \underline{\bar{v}}_{\alpha\beta,s}\right)\right] \cdot \underline{n}_{\alpha\beta} = 0 \quad \text{for phase interfaces within an REV.} \quad (3.38)$$

In this macro-scale formulation, $e_\alpha\left(\rho \eta\right)$ and $\hat{\phi}_\alpha$ are exchange of entropy due to mass transfer and mechanical interactions, respectively.

Proceeding similarly for **interfaces**, Gray and Hassanizadeh [1989] came up with the following substitutions of the general pore-scale balance equation for interfaces,

$$\psi_{\alpha\beta} = \eta_{\alpha\beta}, \quad \underline{i}_{\alpha\beta} = \underline{\phi}_{\alpha\beta}, \quad f_{\alpha\beta} = b_{\alpha\beta}, \quad G_{\alpha\beta} = \Lambda_{\alpha\beta}, \quad \psi_\alpha = \eta_\alpha, \quad \underline{i}_\alpha = \underline{\phi}_\alpha. \quad (3.39)$$

With these substitutions, they came up with the following macro-scale entropy balance equation for interfaces:

$$\frac{\partial}{\partial t}\left(\Gamma_{\alpha\beta}\bar{\eta}_{\alpha\beta}a_{\alpha\beta}\right) + \nabla \cdot \left(\Gamma_{\alpha\beta}\underline{\bar{v}}_{\alpha\beta,s}\bar{\eta}_{\alpha\beta}a_{\alpha\beta}\right) - \nabla \cdot \left(\underline{\phi}_{\alpha\beta}a_{\alpha\beta}\right)$$
$$- \Gamma_{\alpha\beta}\bar{b}_{\alpha\beta}a_{\alpha\beta} - \Gamma_{\alpha\beta}\bar{\Lambda}_{\alpha\beta}a_{\alpha\beta}$$
$$= \sum_{\alpha=1}^{2} \frac{1}{A_{\alpha\beta}}\int_{A_{\alpha\beta}} \left[\rho_\alpha \eta_\alpha\left(\underline{v}_\alpha - \underline{v}_{\alpha\beta}\right) + \hat{\underline{\phi}}_{\alpha\beta}\right] \cdot \underline{n}_{\alpha\beta} \, dA. \quad (3.40)$$

The second law of thermodynamics states that entropy can only be produced or stay constant. Macroscopically, this can be formulated as

$$\sum_\alpha \bar{\Lambda}_\alpha + \sum_{\alpha\beta} \bar{\Lambda}_{\alpha\beta} \geq 0. \quad (3.41)$$

3.3 Constitutive relationships

Using the alternative theory, constitutive relationships do not need to be postulated completely empirically. Instead, Hassanizadeh and Gray [1990] used their set of volume-averaged macro-scale conservation equations to combine them with the second law of thermodynamics,

$$\Lambda = \sum_\alpha \Lambda_\alpha + \sum_{\alpha\beta} \Lambda_{\alpha\beta} \geq 0. \tag{3.42}$$

Note that from now on, considerations will take place on the macro scale. For ease of reading, overbars are left out as it is automatically implied that all parameters are macro-scale parameters. In a next step, Hassanizadeh and Gray [1990] introduced specific Helmholtz free energies for phases, A_α, and for interfaces, $A_{\alpha\beta}$ which are defined by

$$A_\alpha = u_\alpha - T_\alpha \eta_\alpha, \tag{3.43}$$

$$A_{\alpha\beta} = u_{\alpha\beta} - T_{\alpha\beta} \eta_{\alpha\beta}, \tag{3.44}$$

where T_α and $T_{\alpha\beta}$ are the temperatures of phase α, or interface $\alpha\beta$, respectively. Also, the entropy fluxes of phases and interfaces are defined through

$$\nabla \cdot \underline{\phi}_\alpha = \underline{q}_\alpha \nabla T_\alpha, \tag{3.45}$$

$$\nabla \cdot \underline{\phi}_{\alpha\beta} = \underline{q}_{\alpha\beta} \nabla T_{\alpha\beta}, \tag{3.46}$$

where $\underline{q}_\alpha$ is the heat flux of phase α and $\underline{q}_{\alpha\beta}$ that of phase $\alpha\beta$, respectively. Resolving the entropy balance equations, Eq. (3.36) and (3.40), for Λ_α and $\Lambda_{\alpha\beta}$, respectively, and making use of the above expressions for specific Helmholtz free energies and entropy fluxes yields

$$\Lambda_\alpha = -\frac{\phi_\alpha \rho_\alpha}{T_\alpha}\left(\frac{D_\alpha A_\alpha}{Dt} + \eta_\alpha \frac{D_\alpha T_\alpha}{Dt}\right) + \frac{\phi_\alpha}{(T_\alpha)^2}\underline{q}_\alpha \cdot \nabla T_\alpha + \frac{\phi_\alpha}{T_\alpha}\underline{\underline{t}}_\alpha : \nabla \underline{v}_{\alpha,s}$$

$$-\sum_{\beta\neq\alpha}\left(\hat{\Phi}^\alpha_{\alpha\beta} - \frac{\hat{Q}^\alpha_{\alpha\beta}}{T_\alpha}\right) \qquad (3.47)$$

$$\Lambda_{\alpha\beta} = -\frac{a_{\alpha\beta}\Gamma_{\alpha\beta}}{T_{\alpha\beta}}\left(\frac{D_{\alpha\beta}A_{\alpha\beta}}{Dt} + \eta_{\alpha\beta}\frac{D_{\alpha\beta}T_{\alpha\beta}}{Dt}\right) + \frac{a_{\alpha\beta}}{(T_\alpha)^2}\underline{q}_{\alpha\beta} \cdot \nabla T_{\alpha\beta} + \frac{a_{\alpha\beta}}{T_{\alpha\beta}}\underline{\underline{t}}_{\alpha\beta} : \nabla \underline{v}_{\alpha\beta,s}$$

$$+\left\{\hat{\Phi}^\alpha_{\alpha\beta} - \frac{1}{T_{\alpha\beta}}\left[\hat{Q}^\alpha_{\alpha\beta} + \hat{\underline{T}}^\alpha_{\alpha\beta} \cdot \underline{v}_{\alpha,\alpha\beta} + \hat{e}^\alpha_{\alpha\beta}\left(A_{\alpha,\alpha\beta} - \eta_\alpha T^\alpha_{\alpha\beta} + \frac{1}{2}v^2_{\alpha,\alpha\beta}\right)\right]\right\}$$

$$+\left\{\hat{\Phi}^\beta_{\alpha\beta} - \frac{1}{T_{\alpha\beta}}\left[\hat{Q}^\beta_{\alpha\beta} + \hat{\underline{T}}^\beta_{\alpha\beta} \cdot \underline{v}_{\beta,\alpha\beta} + \hat{e}^\beta_{\alpha\beta}\left(A_{\beta,\alpha\beta} - \eta_\beta T^\beta_{\alpha\beta} + \frac{1}{2}v^2_{\beta,\alpha\beta}\right)\right]\right\} \quad (3.48)$$

where $\hat{\Phi}^\alpha_{\alpha\beta}$ is the body supply of entropy from the $\alpha\beta$-interface to the α-phase, $\frac{D_\alpha}{Dt}$ and $\frac{D_{\alpha\beta}}{Dt}$ are the material derivatives following the motion of the α-phase and the $\alpha\beta$-interface, respectively, and finally, $\underline{v}_{\alpha,\alpha\beta}$ as well as $\underline{v}_{\beta,\alpha\beta}$ are the velocities of the α- and β-phase, respectively, with respect to the $\alpha\beta$-interface.

Substituting the modified entropy balance equations, Eq.s (3.47) and (3.48), into the entropy inequality given by Eq. (3.42) leads to the following equation:

$$\sum_\alpha \Lambda_\alpha + \sum_{\alpha\beta}\Lambda_{\alpha\beta} = -\sum_\alpha \frac{\phi_\alpha \rho_\alpha}{T_\alpha}\left(\frac{D_\alpha A_\alpha}{Dt} + \eta_\alpha \frac{D_\alpha T_\alpha}{Dt}\right) + \sum_\alpha \frac{\phi_\alpha}{(T_\alpha)^2}\underline{q}_\alpha \cdot \nabla T_\alpha$$

$$+\sum_\alpha \frac{\phi_\alpha}{T_\alpha}\underline{\underline{t}}_\alpha : \nabla \underline{v}_{\alpha,s} - \sum_{\alpha\beta}\frac{a_{\alpha\beta}\Gamma_{\alpha\beta}}{T_{\alpha\beta}}\left(\frac{D_{\alpha\beta}A_{\alpha\beta}}{Dt} + \eta_{\alpha\beta}\frac{D_{\alpha\beta}T_{\alpha\beta}}{Dt}\right)$$

$$+\sum_{\alpha\beta}\frac{a_{\alpha\beta}}{(T_\alpha)^2}\underline{q}_{\alpha\beta} \cdot \nabla T_{\alpha\beta} + \sum_{\alpha\beta}\frac{a_{\alpha\beta}}{T_{\alpha\beta}}\underline{\underline{t}}_{\alpha\beta} : \nabla \underline{v}_{\alpha\beta,s}$$

$$+\sum_\alpha \sum_{\beta\neq\alpha}\frac{1}{T_{\alpha\beta}}\left[\frac{\hat{Q}^\alpha_{\alpha\beta}T_{\alpha\beta,\alpha}}{T_\alpha} + \hat{\underline{T}}^\alpha_{\alpha\beta} \cdot \underline{v}_{\alpha,\alpha\beta}\right.$$

$$\left.+ \hat{e}^\alpha_{\alpha\beta}\left(A^\alpha_{\alpha\beta} + \frac{1}{2}\left(v^\alpha_{\alpha\beta}\right)^2 + \eta_\alpha T^\alpha_{\alpha\beta}\right)\right] \geq 0 \,. \qquad (3.49)$$

The complete macro-scale equation system consists of more unknowns than equations. Hassanizadeh and Gray [1990] argue that it consists of $\frac{5N(N+1)}{2}$ equations where N is the number of considered phases which may equal two or three if the solid phase is considered as an extra phase. For the system description, they propose the following set of independent variables:

$$(\phi_\alpha \rho_\alpha), \quad (a_{\alpha\beta}\Gamma_{\alpha\beta}), \quad \underline{v}_{\alpha,s}, \quad \underline{v}_{\alpha\beta,s}, \quad T_\alpha, \quad \nabla T_\alpha, \quad T_{\alpha\beta}, \quad \nabla T_{\alpha\beta},$$
$$\nabla(\phi_\alpha \rho_\alpha), \quad \nabla(a_{\alpha\beta}\Gamma_{\alpha\beta}). \quad (3.50)$$

Note that a variable and its gradient can both be primary variables as their magnitude can be chosen independently. Assuming these variables as primary unknowns, all remaining variables must dependent on these primary unknowns. Specifically, these are

$$A_\alpha, \quad A_{\alpha\beta}, \quad \left(\phi_\alpha \underline{t}_\alpha\right), \quad \left(a_{\alpha\beta}\underline{t}_{\alpha\beta}\right), \quad \hat{\underline{T}}^\alpha_{\alpha\beta}, \quad \left(\phi_\alpha \underline{q}_\alpha\right), \quad \hat{Q}^\alpha_{\alpha\beta}, \quad \left(a_{\alpha\beta}\underline{q}_{\alpha\beta}\right), \quad \eta_\alpha,$$
$$\eta_{\alpha\beta}, \quad \hat{e}^\alpha_{\alpha\beta}. \quad (3.51)$$

Hassanizadeh and Gray [1990] could show that this approach is equivalent to assuming the following set of primary unknowns which are actually parameters themselves and do not include products of parameters any more:

$$\rho_\alpha, \quad \Gamma_{\alpha\beta}, \quad \underline{v}_{\alpha,s}, \quad \underline{v}_{\alpha\beta,s}, \quad T_\alpha, \quad \nabla T_\alpha, \quad T_{\alpha\beta}, \quad \nabla T_{\alpha\beta}, \quad \phi, \quad \nabla\phi, \quad S_w,$$
$$\nabla S_w, \quad a_{\alpha\beta}, \quad \nabla a_{\alpha\beta}. \quad (3.52)$$

The secondary unknowns,

$$A_\alpha, \quad A_{\alpha\beta}, \quad \underline{t}_{\alpha\beta}, \quad \underline{t}_{\alpha\beta}, \quad \hat{\underline{T}}^\alpha_{\alpha\beta}, \quad \underline{q}_\alpha, \quad \eta_\alpha, \quad \eta_{\alpha\beta}, \quad \hat{e}^\alpha_{\alpha\beta} \quad (3.53)$$

can then be expressed as functions of the primary unknowns.

In principle, all dependent variables are functions of the complete list of primary variables. However, some of these dependencies are extremely weak and can therefore be neglected. Having this fact in mind and exploiting the facts that constitutive relationships may violate neither conservation laws nor the second law of thermodynamics, Hassanizadeh and Gray [1990] therefore propose the following constitutive dependencies of Helmholtz free energies:

$$A_\alpha = A_\alpha(\rho_\alpha, T_\alpha, S_\alpha) \quad (3.54)$$
$$A_{\alpha\beta} = A_{\alpha\beta}(\Gamma_{\alpha\beta}, T_{\alpha\beta}, a_{\alpha\beta}, S_w) \quad (3.55)$$

Note that it is absolutely required to let Helmholtz free energies of phases or interfaces depend on saturation. If neither of them is a function of saturation the second law of thermodynamics would yield $p_n = p_w$, which is not supported experimentally. Inserting these dependencies along with the thermodynamic definitions of pressures and macroscopic interfacial tension,

$$p_\alpha(\rho_\alpha, T_\alpha, S_w) = \rho_\alpha^2 \frac{\partial A_s}{\partial \rho_s} \tag{3.56}$$

$$\gamma_{\alpha\beta}(\Gamma_{\alpha\beta}, T_{\alpha\beta}, a_{\alpha\beta}, S_w) = -\Gamma_{\alpha\beta}^2 \frac{\partial A_{\alpha\beta}}{\partial \Gamma_{\alpha\beta}} \tag{3.57}$$

$$= -a_{\alpha\beta}\Gamma_{\alpha\beta}\frac{\partial A_{\alpha\beta}}{\partial a_{\alpha\beta}} \tag{3.58}$$

into the entropy inequality leads—after some manipulation—to

$$\sum_\alpha \Lambda_\alpha + \sum_{\alpha\beta} \Lambda_{\alpha\beta} = -\sum_\alpha \frac{\phi\rho_\alpha}{T_\alpha} \frac{D_\alpha T_\alpha}{Dt} \left(\frac{\partial A_\alpha}{\partial T_\alpha} + \eta_\alpha\right)$$

$$-\sum_{\alpha\beta} \frac{a_{\alpha\beta}\Gamma_{\alpha\beta}}{T_{\alpha\beta}} \frac{D_{\alpha\beta} T_{\alpha\beta}}{Dt} \left(\frac{\partial A_{\alpha\beta}}{\partial T_{\alpha\beta}} + \eta_{\alpha\beta}\right)$$

$$+ \sum_{\alpha \neq s} \frac{\underline{\underline{d}}_\alpha}{T_\alpha} \cdot \nabla \cdot \left(\phi S_\alpha \left(p_\alpha \underline{\underline{I}} + \underline{\underline{t}}_\alpha\right)\right)$$

$$+ \sum_{\alpha \neq s} \frac{1}{T_\alpha} \underline{v}_{\alpha,s} \cdot \left(p_\alpha \nabla (\phi S_w) - \phi S_\alpha \frac{\partial A_\alpha}{\partial S_\alpha} \nabla S_\alpha\right)$$

$$- \sum_{\beta \neq \alpha} \hat{\underline{T}}^\alpha_{\alpha\beta} + \sum_{\beta \neq \alpha} \hat{\underline{T}}^\alpha_{\alpha\beta} \frac{T_{\alpha\beta,\alpha}}{T_{\alpha\beta}}$$

$$+ \sum_{\alpha\beta} \frac{1}{T_{\alpha\beta}} \underline{v}_{\alpha\beta,s} \cdot \left(-a_{\alpha\beta}\Gamma_{\alpha\beta} \frac{\partial A_{\alpha\beta}}{\partial S_w} \nabla S_w + \hat{\underline{T}}^\alpha_{\alpha\beta} + \hat{\underline{T}}^\beta_{\alpha\beta}\right)$$

$$- \frac{\partial S_w}{\partial t} \left(\frac{\phi}{T_w} p_w - \frac{\phi}{T_n} p_n - \frac{\phi S_w \rho_w}{T_w} \frac{\partial A_w}{\partial S_w} + \frac{\phi S_n \rho_n}{T_n} \frac{\partial A_n}{\partial S_n}\right)$$

$$- \sum_\alpha \frac{\Gamma_{\alpha\beta} a_{\alpha\beta}}{T_{\alpha\beta}} \frac{\partial A_{\alpha\beta}}{\partial S_w}$$

$$+ \sum_\alpha \sum_{\alpha\beta} \frac{T_{\alpha\beta,\alpha}}{T_{\alpha\beta} T_\alpha} \left(\hat{Q}^\alpha_{\alpha\beta}\right) + \sum_\alpha \nabla T_\alpha \cdot \left(\frac{\phi q_\alpha}{T_\alpha^2}\right) + \sum_{\alpha\beta} \nabla T_{\alpha\beta} \cdot \left(\frac{a_{\alpha\beta} q_{\alpha\beta}}{T_{\alpha\beta}^2}\right)$$

$$+ \sum_\alpha \sum_{\beta \neq \alpha} \frac{\hat{e}^\alpha_{\alpha\beta}}{T_{\alpha\beta}} \left(G_{\alpha\beta,a} + \frac{1}{2} v^2_{\alpha\beta,\alpha}\right)$$

$$+ \sum_\alpha \sum_{\beta \neq \alpha} \frac{\hat{e}^\alpha_{\alpha\beta}}{T_{\alpha\beta}} \left(\eta_\alpha T_{\alpha\beta,\alpha} - \frac{p_\alpha T_{\alpha\beta,\alpha}}{\rho_\alpha T_\alpha}\right) \geq 0. \qquad (3.59)$$

Here, G_α and $G_{\alpha\beta}$ are the Gibbs free energies of phase α or interface $\alpha\beta$, respectively, which are defined by

$$G_\alpha = A_\alpha + \frac{p_\alpha}{\rho_\alpha} \qquad (3.60)$$

$$G_{\alpha\beta} = A_{\alpha\beta} + \gamma_{\alpha\beta} \Gamma_{\alpha\beta}, \qquad (3.61)$$

$\underline{\underline{d}}_\alpha$ is the deformation rate tensor, $T_{\alpha,\beta}$ and $T_{\alpha\beta,\alpha}$ denote relative temperatures (temperature differences) between the phases / interfaces given in the subscript. In the above form of the residual entropy inequality, $\underline{\underline{t}}_{\alpha\beta} = \gamma_{\alpha\beta}\underline{\underline{I}}$ was directly taken into account.

It is now possible to determine and investigate the state of local thermodynamic equilibrium with respect to the entropy inequality and extract constitutive dependencies. At local thermodynamic equilibrium, the variables

$$\underline{v}_{\alpha,s}, \quad \underline{v}_{\alpha\beta,s}, \quad T_{\alpha,s}, \quad T_{\alpha\beta,\alpha}, \quad \nabla T_\alpha, \quad \nabla T_{\alpha\beta}, \quad \frac{DS_w}{Dt}, \quad \hat{e}^\alpha_{\alpha\beta} \qquad (3.62)$$

vanish. These conditions imply that at equilibrium, there will be no relative movement of phases and interfaces; the rate of change of saturations and all temperature differences, both temperature gradients and temperature differences between phases and interfaces, will be zero. Additionally, the rate of mass exchange between phases and interfaces will be zero.

One can readily verify that at that state, Λ is zero which means that it attains its absolute minimum to Eq. (3.59). For Λ to attain a minimum, the necessary condition is that the derivative of Λ with respect to each of the parameters it depends on ($\underline{v}_{\alpha,s}, \underline{v}_{\alpha\beta,s}, \frac{\partial S_w}{\partial t}, T_{\alpha\beta,\alpha}, \nabla T_\alpha, \nabla T_{\alpha\beta}$, and $\hat{e}^\alpha_{\alpha\beta}$) must become zero at equilibrium. Note that the factors in front of the first three brackets are not considered here; this is due to the fact that our choice of constitutive relationships will make those terms vanhish at all times. But then, each of the terms in brackets in Eq. (3.59) must become zero independently. From this condition, it is easily possible to extract constitutive relationships:

$$\eta_\alpha = -\frac{\partial A_\alpha}{\partial T_\alpha} \qquad (3.63)$$

$$\eta_{\alpha\beta} = -\frac{\partial A_{\alpha\beta}}{\partial T_{\alpha\beta}} \qquad (3.64)$$

$$\underline{\underline{t}}_\alpha = -p_\alpha \underline{\underline{I}} \qquad (3.65)$$

$$(p_n - p_w)_e = p_c \qquad (3.66)$$

$$\left(\hat{\underline{T}}_\alpha\right)_e = \left(\sum_{\beta \neq \alpha} \hat{\underline{T}}^{\alpha\beta}_\alpha\right)_e = p_\alpha \nabla(\phi S_\alpha) - \phi S_\alpha \rho_\alpha \frac{\partial A_\alpha}{\partial S_\alpha} \nabla S_\alpha \qquad (3.67)$$

$$\left(\hat{\underline{T}}^\alpha_{\alpha\beta} + \hat{\underline{T}}^\beta_{\alpha\beta}\right)_e = \gamma_{\alpha\beta} \nabla a_{\alpha\beta} - a_{\alpha\beta}\Gamma_{\alpha\beta} \frac{\partial A_{\alpha\beta}}{\partial S_w} \nabla S_w \qquad (3.68)$$

where the lower index e indicates evaluation at equilibrium. A definition of macroscale capillary pressure can be extracted from the expression following the $\frac{\partial S_w}{\partial t}$ term in the residual entropy inequality given in Eq. (3.59):

$$\frac{\phi p_c}{T_{wn}} = -\frac{\phi S_w \rho_w}{T_w}\frac{\partial A_w}{\partial S_w} + \frac{\phi S_n \rho_n}{T_n}\frac{\partial A_n}{\partial S_n} - \sum_{\alpha\beta}\frac{a_{\alpha\beta}\Gamma_{\alpha\beta}}{T_{\alpha\beta}}\frac{\partial A_{\alpha\beta}}{\partial S_w}. \qquad (3.69)$$

The interpretation of Eq. (3.69) will be discussed in further detail in Sec. 3.3.2. From Eq.s (3.67) and (3.68), Hassanizadeh and Gray [1990] concluded that momentum exchange is made up of an equilibrium and a non-equilibrium part where the latter vanishes at equilibrium. Specifically, they obtain the following relationships for phases and interfaces,

$$\hat{T}_\alpha = \sum_{\beta\neq\alpha}\hat{T}_{\alpha\beta} = \hat{\bar{T}}_\alpha + \left(\sum_{\beta\neq\alpha}\hat{T}_{\alpha\beta}\right)_e \qquad (3.70)$$

$$-\hat{T}^\alpha_{\alpha\beta} - \hat{T}^\beta_{\alpha\beta} = \hat{T}_{\alpha\beta} = \hat{\bar{T}}_{\alpha\beta} + \left(-\hat{T}^\alpha_{\alpha\beta} - \hat{T}^\beta_{\alpha\beta}\right)_e \qquad (3.71)$$

where $\hat{\bar{T}}_\alpha$ and $\hat{\bar{T}}_{\alpha\beta}$ are the non-equilibrium parts of momentum transfer and are allowed to be general functions of the full set of primary variables.

For the following considerations, a simplified form of the residual entropy inequality is considered that is applicable in many two-phase flow situations:

- local thermal equilibrium. This means that the temperatures of all phases and interfaces within an REV are the same and denoted by

$$T := T_w = T_n = T_s = T_{wn} = T_{ws} = T_{ns}. \qquad (3.72)$$

- immiscible two-phase flow. This means that the mass transfer rates $\hat{e}^{\alpha\beta}_\alpha$ drop out.

After multiplication by T these assumptions lead to a much simpler form of the residual entropy inequality,

$$T\Lambda = \sum_{\alpha \neq s} \underline{v}_{\alpha,s} \cdot \hat{\underline{t}}_\alpha - \sum_{\alpha\beta} \underline{v}_{\alpha\beta} \cdot \hat{\underline{t}}_{\alpha\beta} - \phi \frac{\partial S_w}{\partial t}(p_n - p_w - p_c)$$

$$+ \frac{\nabla T}{T} \cdot \left(\sum_\alpha \phi_\alpha \underline{q}_\alpha + \sum_{\alpha\beta} a_{\alpha\beta} \underline{q}_{\alpha\beta} \right) \geq 0. \quad (3.73)$$

3.3.1 Relative permeability

In this section, we make use of the general momentum balance equation along with the derived constitutive dependencies in order to study the physical meaning of relative permability as a constitutive function of the classical theory. This is done by comparing the extended Darcy's law to the alternative theory.

In the following, we consider a simplification of the general macro-scale phase and interface momentum balance equations given in Eq.s (3.21) and (3.25) where immiscible and incompressible two-phase flow in a rigid porous medium is considered. This is not a necessary assumption, but done in order to focus on the most relevant aspects. These assumptions lead to

$$-\nabla \left(\phi S_\alpha \underline{\underline{t}}_\alpha \right) - \phi \rho_\alpha S_\alpha \underline{g} = \hat{\underline{t}}_\alpha \quad (3.74)$$

$$-\nabla \left(a_{\alpha\beta} \underline{\underline{t}}_{\alpha\beta} \right) - a_{\alpha\beta} \Gamma_{\alpha\beta} \underline{g} = -\hat{\underline{t}}_{\alpha\beta}^\alpha - \hat{\underline{t}}_{\alpha\beta}^\beta. \quad (3.75)$$

Inserting the previously determined constitutive dependencies for $\underline{\underline{t}}_\alpha$, $\hat{\underline{t}}_\alpha$, $\underline{\underline{t}}_{\alpha\beta}$, and $\hat{\underline{t}}_{\alpha\beta}^\alpha$ into the above macro-scale momentum balance equations results in

$$\nabla p_\alpha - \rho_\alpha \underline{g} = -\rho_\alpha \frac{\partial A_\alpha}{\partial S_\alpha} \nabla S_\alpha + \frac{\hat{\underline{t}}_\alpha}{\phi S_\alpha} \quad (3.76)$$

$$\nabla \left(a_{\alpha\beta} \gamma_{\alpha\beta} \right) + \Gamma_{\alpha\beta} a_{\alpha\beta} \underline{g} = \Gamma_{\alpha\beta} a_{\alpha\beta} \frac{\partial A_{\alpha\beta}}{\partial S_w} \nabla S_w - \hat{\underline{t}}_{\alpha\beta}. \quad (3.77)$$

The dissipative drag forces $\hat{\underline{t}}_\alpha$ and $\hat{\underline{t}}_{\alpha\beta}$ are general functions of the primary variables. But for isothermal systems or systems with small effects of temperature gradients that are close to equilibrium, a linear dependence on the primary variables $\underline{v}_{\alpha,s}$ and

$\underline{v}_{\alpha\beta,s}$ can be assumed. This results in

$$\hat{\underline{t}}_\alpha = -\underline{\underline{R}}^n_\alpha \cdot \underline{v}_{n,s} - \underline{\underline{R}}^w_\alpha \underline{v}_{w,s} - \sum_{\beta\gamma} \underline{\underline{R}}^{\beta\gamma}_\alpha \cdot \underline{v}_{\beta\gamma,s} \quad (3.78)$$

$$\hat{\underline{t}}_{\alpha\beta} = -\underline{\underline{R}}^n_{\alpha\beta} \cdot \underline{v}_{n,s} - \underline{\underline{R}}^w_{\alpha\beta} \underline{v}_{w,s} - \sum_{\gamma\delta} \underline{\underline{R}}^{\gamma\delta}_{\alpha\beta} \cdot \underline{v}_{\gamma\delta,s}, \quad (3.79)$$

where the coefficients $\underline{\underline{R}}_\alpha$ and $\underline{\underline{R}}_{\alpha\beta}$ are resistance tensors to flow of phases and interfaces, respectively.

Inserting the linearized relations, Eq.s (3.78) and (3.79), into the simplified momentum balance equations, Eq.s (3.76) and (3.77) results in the following momentum balances for phases and interfaces,

$$\frac{1}{\phi S_w}\left[\sum_{\beta\neq s}\underline{\underline{R}}^\beta_\alpha \cdot \underline{v}_{\beta,s} + \sum_{\beta\gamma}\underline{\underline{R}}^{\beta\gamma}_\alpha \cdot \underline{v}_{\beta\gamma,s}\right] = -(\nabla p_\alpha - \rho_\alpha \underline{g}) - \rho_\alpha \frac{\partial A_\alpha}{\partial S_\alpha}\nabla S_\alpha \quad (3.80)$$

$$\left[\sum_{\gamma\neq s}\underline{\underline{R}}^\gamma_{\alpha\beta} \cdot \underline{v}_{\gamma,s} + \sum_{\gamma\delta}\underline{\underline{R}}^{\gamma\delta}_{\alpha\beta} \cdot \underline{v}_{\gamma\delta,s}\right] = \nabla(a_{\alpha\beta}\gamma_{\alpha\beta}) + a_{\alpha\beta}\Gamma_{\alpha\beta}\underline{g}$$

$$-\Gamma_{\alpha\beta}a_{\alpha\beta}\frac{\partial A_{\alpha\beta}}{\partial S_w}\nabla S_w \quad (3.81)$$

These equations indicate that, if cross-coupling terms are neglected, then movement of phases and interfaces can be described by

$$\frac{1}{\phi S_\alpha}\underline{\underline{R}}^\alpha_\alpha \cdot \underline{v}_{\alpha,s} = -(\nabla p_\alpha - \rho_\alpha \underline{g}) - \rho_\alpha \frac{\partial A_\alpha}{\partial S_\alpha}\nabla S_\alpha \quad (3.82)$$

$$\underline{\underline{R}}^{\alpha\beta}_{\alpha\beta} \cdot \underline{v}_{\alpha\beta,s} = \nabla(a_{\alpha\beta}\gamma_{\alpha\beta}) + a_{\alpha\beta}\Gamma_{\alpha\beta}\underline{g} - \Gamma_{\alpha\beta}a_{\alpha\beta}\frac{\partial A_{\alpha\beta}}{\partial S_w}\nabla S_w. \quad (3.83)$$

These equations provide possibilities to calculate macro-scale velocities, both for phases and interfaces. Concerning interfaces, the alternative approach provides a means for calculating macro-scale interface velocities which was impossible in the framework of the standard approach. Concerning phases it can be seen that Eq. (3.82) differs from the extended Darcy's law by the term $\rho_\alpha \frac{\partial A_\alpha}{\partial S_\alpha}\nabla S_\alpha$. This term may be significant. It can only be neglected if

$$\phi S_\alpha \rho_\alpha \left|\frac{\partial A_\alpha}{\partial S_\alpha}\nabla S_\alpha\right| \ll \left|\underline{\underline{R}}^\alpha_\alpha \cdot \underline{v}_{\alpha,s}\right|. \quad (3.84)$$

This condition is not necessarily fulfilled as close to equilibrium, the velocity $\underline{v}_{\alpha,s}$ may be very small. This means that there may be additional terms in the two-phase version of Darcy's law which are not captured by the classical approach or which are lumped into the relative permeability–saturation functions.

It remains to evaluate the actual interpretation of relative permeability as an empirical factor in classical two-phase flow models in the alternative model framework. Resolving Eq. (3.82) for $\underline{v}_{\alpha,s}$ yields

$$\underline{v}_{\alpha,s} = -\phi S_\alpha \left(\underline{\underline{R}}^\alpha\right)^{-1} \left(\nabla p_\alpha - \rho_\alpha \underline{g} + \phi \rho_\alpha S_\alpha \frac{\partial A_\alpha}{\partial S_\alpha} \nabla S_\alpha\right). \tag{3.85}$$

To maintain correct dimensions and to fulfill the single-phase Darcy's law in the limit a possible definition of $\underline{\underline{R}}^\alpha$ can be obtained,

$$\underline{\underline{R}}^\alpha = \mu_\alpha \phi^2 \underline{\underline{K}}^{-1}. \tag{3.86}$$

Strictly speaking, however, $\underline{\underline{R}}^\alpha$ may still depend on the primary variables. The velocity $\underline{v}_{\alpha,s}$ is related to the Darcy velocity by

$$\underline{v}_\alpha = \frac{\underline{v}_{\alpha,s}}{\phi S_\alpha} \tag{3.87}$$

Inserting Eq.s (3.86) and (3.87) into Eq. (3.85) results in

$$\underline{v}_\alpha = -\frac{S_\alpha^2}{\mu_\alpha} \underline{\underline{K}} \left(\nabla p_\alpha - \rho_\alpha \underline{g} + \rho_\alpha \frac{\partial A_\alpha}{\partial S_\alpha} \nabla S_\alpha\right). \tag{3.88}$$

Comparing Eq. (3.88) with the classical extended Darcy's law and considering Eq. (3.86) suggests that a natural choice of the relative permeability functions in the classical approach is given by

$$k_{r\alpha} = S_\alpha^2, \tag{3.89}$$

although we have to keep in mind that this result only comes due to the assumption of $\underline{\underline{R}}^\alpha$ being constant. This choice of relative permeability–saturation relationship, however, is only valid if the constraint given in Eq. (3.84) is fulfilled which is not necessarily the case. It follows that the effects of the factor S_α^2 and the term $\rho_\alpha \frac{\partial A_\alpha}{\partial S_\alpha} \nabla S_\alpha$ are lumped into relative permeability. In other words, the relative permeability–saturation relationship in the classical model deviates from quadratic dependency on saturation due to the neglect of the term $\rho_\alpha \frac{\partial A_\alpha}{\partial S_\alpha} \nabla S_\alpha$ (and possibly due to nonlinearity of $\underline{\underline{R}}^\alpha_\alpha$).

Side remarks: The assumption of a linear theory close to equilibrium gives further insights, e.g. with respect to capillary pressure, heat flux, and driving forces in the macro-scale momentum balance:

- From the residual entropy inequality (Eq. (3.73)) it can be deduced that the time rate of change of saturation, $\frac{\partial S_w}{\partial t}$ can be expressed as a function of bulk fluid phase pressures and capillary pressure as

$$\frac{\partial S_w}{\partial t} = -\Pi_w \left(p_n - p_w - p_c\right), \quad (3.90)$$

where Π_w is a general functional which must be determined experimentally. It might, however, be a function of ρ_α, ϕ, S_w, $a_{\alpha\beta}$, $\Gamma_{\alpha\beta}$, and T.

- Linearizing the heat flux occuring in the entropy inequality given in Eq. (3.73) leads to

$$\sum_\alpha \phi_\alpha \underline{q}_\alpha + \sum_{\alpha\beta} a_{\alpha\beta} \underline{q}_{\alpha\beta} = \lambda \nabla T, \quad (3.91)$$

where λ can be identified as the thermal conductivity of the medium.

- From Eq. (3.82), another interesting form of the momentum balance equation for phases and interfaces can be deduced,

$$\underline{\underline{R}}^\alpha_\alpha \cdot \underline{v}_{\alpha,s} = -\phi S_\alpha \rho_\alpha \left(\nabla G_\alpha - \underline{g}\right) \quad (3.92)$$

$$\underline{\underline{R}}^{\alpha\beta}_{\alpha\beta} \cdot \underline{v}_{\alpha\beta,s} = -a_{\alpha\beta} \Gamma_{\alpha\beta} \left(\nabla G_{\alpha\beta} - \underline{g}\right), \quad (3.93)$$

where G_α and $G_{\alpha\beta}$ are the specific Gibbs free energies of phases and interfaces, respectively. They are related to specific Helmholtz free energies through

$$G_\alpha = A_\alpha + \frac{p_\alpha}{\rho_\alpha} \quad (3.94)$$

$$G_{\alpha\beta} = A_{\alpha\beta} + \frac{\gamma_{\alpha\beta}}{\Gamma_{\alpha\beta}}. \quad (3.95)$$

This form of the momentum balance equation indicates that flow of a phase or interface is driven by the gradient of specific Gibbs free energy of that phase or interface and by gravity.

3.3.2 Capillary pressure

In this section, we start revisiting the insights that the rational thermodynamics approach of Hassanizadeh and Gray provided with respect to capillary pressure

1. From Eq. (3.69) it follows that in the case of local thermodynamic equilibrium, macro-scale capillary pressure can be defined thermodynamically by

$$p_c = -S_w \rho_w \frac{\partial A_w}{\partial S_w} + S_n \rho_n \frac{\partial A_n}{\partial S_w} - \sum_{\alpha\beta} \frac{a_{\alpha\beta}\Gamma_{\alpha\beta}}{\phi} \frac{\partial A_{\alpha\beta}}{\partial S_w} \qquad (3.96)$$

2. A definition of macro-scale phase pressures is provided by Eq. (3.56),

$$p_\alpha = \rho_\alpha^2 \frac{\partial A_\alpha}{\partial \rho_\alpha}. \qquad (3.97)$$

3. Finally, based on the residual entropy inequality and the assumption of a linear theory close to equilibrium, it was found in Eq. (3.90), that a relation between the above variables must exist of the form

$$\frac{\partial S_w}{\partial t} = -\Pi_w \left(p_n - p_w - p_c \right), \qquad (3.98)$$

where Π_w is a non-negative material parameter.

Let us proceed by considering the last equation which defines the relation between p_c and $p_n - p_w$. It can be clearly seen that $p_c = p_n - p_w$ is valid at equilibrium, i.e. in case $\frac{\partial S_w}{\partial t}$ vanishes. If that is not the case, the parameter Π_w plays an essential role. It may be interpreted as a measure for the speed at which the system reacts to a disturbance from equilibrium, e.g. a change in saturation or in $p_n - p_w$. If Π_w is very large, than $p_c = p_n - p_w$ will be reestablished very quickly after the equilibrium state is disturbed. Thus, the relation $p_c = p_n - p_w$ that is assumed by the classical two-phase flow theory is intrinsically assuming that either the system dynamics are such that any disturbance from the equilibrium state is instantaneously eliminated ($\Pi_w \gg 0$) or that the system is assumed to be always in equilibrium, $\frac{\partial S_w}{\partial t} = 0$. The latter is, of course, not the situation which is commonly of interest. Instead, the movement of phases and interfaces, i.e. a non-equilibrium situation where $\frac{\partial S_w}{\partial t} \neq 0$ is the interesting case. Therefore, when using the classical relationship $p_c = p_n - p_w$, it is necessary to ensure that the dynamics of the system are negligible.

By trend, equilibrium is quickly attained if permeability is high. However, dynamics are important if permeability is low. Therefore, an application of the classical approach to low-permeable media appears questionable. After these considerations, it can be concluded that the classically assumed relation, $p_c = p_n - p_w$ is not uniquely valid. Instead, in the general case, the dynamics of the system have to be accounted for which comes into play by a factor Π_w. This issue has been observed experimentally by measuring $p_n - p_w$ under static and dynamic conditions by Stauffer [1978], DiCarlo [2004], Bottero et al. [2006], investigated theoretically by Hassanizadeh and Gray [1990, 1993b] and studied numerically on pore (Mirzaei and Das [2007]) and macro scale (DiCarlo [2005], Manthey et al. [2005]). In the theoretical and numerical works, a formula is applied which varies the classical formula for p_c by an additional dynamic term,

$$p_c = p_n - p_w - \tau \frac{\partial S_w}{\partial t}. \tag{3.99}$$

The parameter τ may depend on the whole list of primary variables. So far, it has been shown that it significantly depends on saturation S_w.

The relation given in Eq. (3.99) provides a definition of macro-scale capillary pressure depending on the bulk phase pressures, but it is not sufficient for a complete description of the system. In the system of macro-scale balance equations there are still more unknown variables than equations. Therefore, the system of macro-scale balance equations needs to be closed by the formulation of constitutive equations, also for capillary pressure. In principle, capillary pressure is a function of the whole list of identified primary variables. For many cases of practical interest, however, this can be simplified. If phase densities, interfacial mass densities, temperature and porosity are assumed to be constant, then a dependence

$$p_c = p_c\left(S_w, a_{wn}, a_{ns}, a_{ws}\right) \tag{3.100}$$

is suggested by Eq. (3.96). If one further considers the fact that the sum of the fluid–solid interfacial areas, $a_{ws} + a_{ns}$ is (practically) constant as all of the solid surface is either covered by wetting phase or by non-wetting phase, then, the combined impact of the fluid–solid interfacial areas can be neglected and a simpler, but still rather general functional form of the constitutive dependence of capillary pressure remains,

$$p_c = p_c\left(S_w, a_{wn}\right). \tag{3.101}$$

This relationship suggests that capillary pressure is not only a function of saturation, but additionally dependent on specific fluid–fluid interfacial area. Hassanizadeh and Gray [1993b] have suggested that the non-uniqueness in the classical capillary pressure–saturation relationship is due to exactly this absence of specific interfacial area in the capillary pressure–saturation relationship. They conjectured that the relationship p_c–S_w–a_{wn} will be unique and independent of the history of flow. This conjecture has been investigated both experimentally and numerically. Experimentally, Brusseau et al. [1997], Chen and Kibbey [2006], Culligan et al. [2004], Schaefer et al. [2000], Wildenschild et al. [2002]. Chen et al. [2007] even showed that using $p_c(S_w, a_{wn})$ relationships, hysteresis can indeed be eliminated within the bounds of the measurement error. Reeves and Celia [1996] calculated specific interfacial area–capillary pressure–saturation surfaces, both for drainage and imbibition. While their model gave a difference between these two surfaces, Held and Celia [2001] showed that including effects of snap-off and local fluid configurations during imbibition, the difference between the two surfaces can be reduced down to a very small value. Finally, Joekar-Niasar et al. [2008] studied different trapping assumptions and obtained both interfacial area–capillary pressure–saturation surfaces and interfacial area–relative permeability–saturation surfaces. They found that imbibition and drainage surfaces had an average difference of 7%. Nuske [2009] applied a sort of pore-network model (percolation model) to fractures in order to obtain $p_c(S_w, a_{wn})$ relations. He discretized a fracture by a raster element model and interpreted each raster element as a pore body. Note that generally, the relation between specific interfacial area, capillary pressure, and saturation is formulated as $a_{wn}(S_w, p_c)$ instead of $p_c(S_w, a_{wn})$. This is due to the fact that only the relationship $a_{wn}(S_w, p_c)$ is single-valued.

Unlike the classical capillary pressure–saturation relationship, the $a_{wn}(S_w, p_c)$ function represents a three-dimensional surface. There is of course a relationship between the $a_{wn}(S_w, p_c)$ surface and the capillary pressure–saturation hysteresis loop observed in classical models: when projecting the $a_{wn}(S_w, p_c)$ data points onto the $p_c(S_w)$ plain, the classical hysteretic capillary pressure saturation relationship can be obtained.

An exemplary $a_{wn}(S_w, p_c)$ surface is shown in Fig. 3.7 (taken from Nuske [2009]). The contour levels indicate specific interfacial areas. Also, the projection of scanning curves onto the capillary pressure–saturation plane is shown.

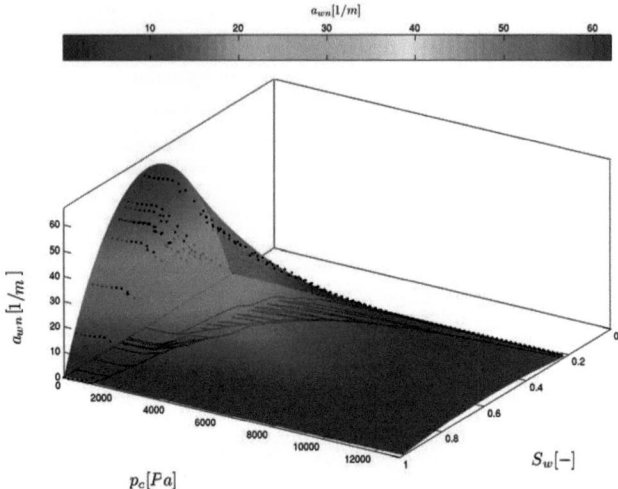

Figure 3.7: Exemplary $a_{wn}(S_w, p_c)$ surface with scanning curves in the $p_c(S_w)$ plane (taken from Nuske [2009]).

It remains to discuss parameterizations of the $a_{wn}(S_w, p_c)$ surface, similar to the previously presented parameterizations of Brooks & Corey or van Genuchten for the classical $p_c(S_w)$ relationships.

Joekar-Niasar et al. [2008] proposed a bi-quadratic dependence of a_{wn} on S_w and p_c,

$$a_{wn}(S_w, p_c) = a_{00} + a_{10}S_w + a_{01}p_c + a_{11}S_w p_c + a_{20}S_w^2 + a_{02}p_c^2. \quad (3.102)$$

Using this parametrization with the six fitting parameters a_{ij}, Joekar-Niasar et al. [2008] obtained fitting coefficients > 0.95 for their data points from a static pore-network model.

An alternative parametrization was proposed by Nuske [2009],

$$a_{wn}(S_w, p_c) = a_1 \cdot (S_{wr} - S_w) \cdot (1 - S_w) + a_2 \cdot (S_{wr} - S_w) \cdot (1 - S_w) \cdot e^{a_3 \cdot p_c}, \quad (3.103)$$

which takes the residual wetting-phase saturation into account. This formulation has several advantages. First of all, it needs less coefficients for obtaining a generally better fit of the data (three instead of six coefficients, but higher correlation coefficients). Secondly, this formulation naturally fulfills the physical constraint that the specific interfacial area is zero if there is only one (mobile) phase – that is for $S_w = 1$ or $S_w = S_{wr}$. Another advantageous property of this parameterization is the numerical behavior. While the polynomial fitting function given in Eq. (3.102) may cause numerical problems due to large gradients the parametrization given in Eq. (3.103) seems to be better behaved in this respect, see Faigle [2009].

Let us briefly recall the main findings of this section:

1. Capillary pressure is only equal to the difference in bulk phase pressures, $p_n - p_w$, at equilibrium (i.e. no flow, $\frac{\partial S_w}{\partial t} = 0$) or can be approximated in that way if dynamic effects are negligible (very small value of τ or very large values of $\underline{K}$). If neither of these conditions is fulfilled the classical model provides wrong results.

2. There is a constitutive relationship between capillary pressure, specific interfacial area, and saturation. The neglect of specific interfacial area in this relationship leads to hysteresis in the classical model. Parameterizations of the $a_{wn}(S_w, p_c)$ surface have been proposed that provide high fitting coefficients.

3.4 Interphase mass transfer

Interphase mass transfer is an issue which needs detailed discussion, although mass transfer terms have already been included in the mass balance equations. Two issues need special attention:

1. When interface mass transfer occurs it is essential to formulate balance equations separately for the different components: for each component in each phase, a separate balance equation needs to be formulated.

2. The interphase mass transfer terms in the macro-scale balance equations are still defined on the pore scale. A macro-scale closure for these equations has to be provided.

Let us first study the appropriate balance equations. Niessner and Hassanizadeh [2009a] extended the set of balance equations derived by Hassanizadeh and Gray [1979b] and Gray and Hassanizadeh [1989, 1998] for the case of flow of two pure fluid phases with no mass transfer to the case of two fluid phases, each made of two components and obtained the following equations:

mass balance for phase components:

$$\frac{\partial \left(\phi S_w \bar{\rho}_w \bar{X}_w^\kappa\right)}{\partial t} + \nabla \cdot \left(\phi S_w \bar{\rho}_w \bar{X}_w^\kappa \bar{\underline{v}}_{w,s}\right) + \nabla \cdot \left(\phi S_w \bar{\underline{j}}_w^\kappa\right)$$
$$= \bar{\rho}_w Q_w^\kappa + \frac{1}{V} \int_{A_{wn}} \left[\rho_w X_w^\kappa \left(\underline{v}_{wn} - \underline{v}_w\right) - \underline{j}_w^\kappa\right] \cdot \underline{n}_{wn} \, dA \quad (3.104)$$

$$\frac{\partial \left(\phi S_n \bar{\rho}_n \bar{X}_n^\kappa\right)}{\partial t} + \nabla \cdot \left(\phi S_n \bar{\rho}_n \bar{X}_n^\kappa \bar{\underline{v}}_{n,s}\right) + \nabla \cdot \left(\phi S_n \bar{\underline{j}}_n^\kappa\right)$$
$$= \bar{\rho}_n Q_n^\kappa + \frac{1}{V} \int_{A_{wn}} \left[\rho_n X_n^\kappa \left(\underline{v}_n - \underline{v}_{wn}\right) + \underline{j}_n^\kappa\right] \cdot \underline{n}_{wn} \, dA \quad (3.105)$$

mass balance for wn-interface components:

$$\frac{\partial \left(\bar{\Gamma}_{wn} \bar{X}_{wn}^\kappa a_{wn}\right)}{\partial t} + \nabla \cdot \left(\bar{\Gamma}_{wn} \bar{X}_{wn}^\kappa a_{wn} \bar{\underline{v}}_{wn,s}\right) + \nabla \cdot \left(\bar{\underline{j}}_{wn}^\kappa a_{wn}\right)$$
$$= \frac{1}{V} \int_{A_{wn}} \left[\rho_w X_w^\kappa \left(\underline{v}_w - \underline{v}_{wn}\right) + \underline{j}_w^\kappa - \rho_n X_n^\kappa \left(\underline{v}_n - \underline{v}_{wn}\right) - \underline{j}_n^\kappa\right] \cdot \underline{n}_{wn} \, dA \quad (3.106)$$

It remains to specify macro-scale closure relationships for the **mass transfer terms** on the right hand sides. The difference between phase and interface velocity directly at the interface is in most cases very small and therefore neglected. This leaves the challenge of finding a macro-scale closure for the diffusive mass transfer term. This term is commonly described by Fick's law and then expanded to first order. Thus, the following formula is obtained:

$$\underline{j}_\alpha^\kappa \cdot \underline{n}_{wn} = \pm D^\kappa \nabla X_\alpha^\kappa \cdot \underline{n}_{wn} \quad (3.107)$$
$$\approx \pm \frac{\rho_\alpha D^\kappa}{d^\kappa} a_{wn} \left(X_{\alpha,s}^\kappa - X_\alpha^\kappa\right), \quad (3.108)$$

where D^κ is the micro-scale Fickian diffusion coefficient for component κ, d^κ is the diffusion length of component κ, $X_{\alpha,s}^\kappa$ is the solubility limit of component κ in phase α, and X_α^κ is the micro-scale mass fraction of component κ in phase α at a distance d^κ away from the interface. It can be clearly seen that the mass transfer rate is proportional to the fluid–fluid specific interfacial area. The variable which is called solubility limit here represents the mass fraction which corresponds to local chemical equilibrium. Its value can be obtained by using equilibrium relationships, such as Henry's law or Raoult's law, see Sec. 2.4.4.

A clever choice of the diffusion length is such that the pore-scale mass fraction X_α^κ in a distance of d^κ from the interface corresponds to the mass fraction $\bar{X}_\alpha^\kappa$ of the considered REV. With that choice of d^κ, Eq. (3.108) can be written as

$$\underline{j}_\alpha^\kappa \cdot \underline{n}_{wn} \approx \pm \frac{\rho_\alpha D^\kappa}{d^\kappa} a_{wn} \left(X_{\alpha,s}^\kappa - \bar{X}_\alpha^\kappa \right) \tag{3.109}$$

With this closure relationship, it is possible to model interphase mass transfer in a physically based way. While in the classical model it was necessary to postulate local chemical equilbrium (unless of course, an empirical description of interphase mass transfer is used) the interfacial-area-based model offers a natural way to account for local chemical non-equilibrium leading to kinetic interphase mass transfer.

3.5 Interphase energy transfer

For interphase energy transfer, similar considerations as for mass transfer can be made, but with respect to the energy balance equation. When accounting for interphase energy transfer the assumption of local thermal equilibrium that was essential in the classical model can be relaxed. This means that the phase temperatures within an REV are not necessarily equal ($T = T_w = T_n = T_s$ no longer has to be assumed). But then, an energy balance equation has to be formulated for each phase (fluid phases and solid phase). A major difference of interphase energy transfer from interphase mass transfer is the fact that energy transfer does not only take place across fluid–fluid interfaces, but additionally across fluid–solid interfaces. Fig. 3.8 illustrates this situation: on the pore scale, mass is transferred across the fluid–fluid interface and energy is transferred between all three phases. Phases are denoted by α and may be the wetting fluid phase (index w), the non-wetting fluid phase (index n), or the solid phase (index s). On the pore scale, temperature and composition of phases may vary continuously throughout the domain. In the classical approach, the procedure is to average this pore-scale situation over a representative elementary volume (REV) yielding porosity, phase saturations, and one average temperature T. Also, if empirical models are to be avoided, local equilibrium between phases with respect to mass transfer has to be assumed. This means that equilibrium relationships, such as Henry's law and Raoult's law, are used. Thus,

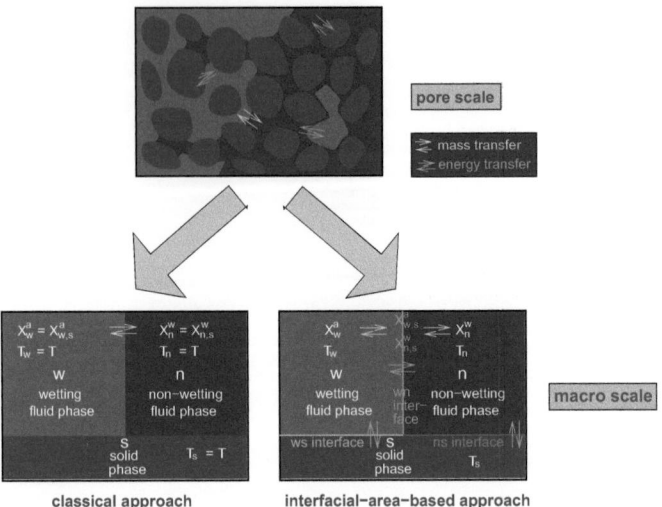

Figure 3.8: Conceptual models with respect to heat and energy transfer on the pore and macro scale using a classical approach and using an interfacial-area-based model.

mass fractions for component κ in phase α, X_α^κ, that correspond to the equilibrium values $X_{\alpha,s}^\kappa$ are determined. A model using interfacial areas could yield mass fractions within the phases that do not correspond to the equilibrium values, or allow for different temperatures T_α for different phases.

Recently, non-equilibrium effects were taken into account in a theoretical approach by Duval et al. (2004) who upscaled pore-scale energy balance equations to the macro scale. However, in order to solve their macro-scale problem they had to obtain a closure by the solution of local pore-scale problems on simple unit cells of known geometry. Also, the previously mentioned approach by Crone et al. [2002] presents a step in the direction of non-equilibrium energy transfer. They developed an extension of the classical model in which they assumed thermal equilibrium between the fluid phases, but non-equilibrium between the fluid mixture and solid. Therefore, they formulated two energy balances, one for the fluid mixture and one for the solid phase. In these, however, interfacial area was not a model variable. Niessner and Hassanizadeh [2009c] developed the first macro-scale model based on the alternative theory which could account for kinetic interphase energy transfer (thermal non-equilibrium) based on interfacial areas. In the following, the procedure developed by Niessner and Hassanizadeh [2009c] is described.

The upscaling of all balance equations from the pore scale to the macro-scale has been discussed in this chapter, but at a certain point, local thermal equilibrium has been assumed. Therefore, the macro-scale energy balance equations will be derived for the case where this assumption is relaxed.

Consider the pore-scale energy balance equations given by

$$\frac{\partial (\rho_n h_n)}{\partial t} + \nabla \cdot (\rho_n h_n \underline{v}_n) = \nabla \cdot (\lambda_n \nabla T_n) \quad (3.110)$$

$$\frac{\partial (\rho_w h_w)}{\partial t} + \nabla \cdot (\rho_w h_w \underline{v}_w) = \nabla \cdot (\lambda_w \nabla T_w) \quad (3.111)$$

$$\frac{\partial (\rho_s h_s)}{\partial t} = \nabla \cdot (\lambda_s \nabla T_s). \quad (3.112)$$

where λ_α is the thermal conductivity of phase α.

In order to fulfill interface equilibrium as defined in the introductory part of Chap. 2, the following conditions must hold at the three interfaces $A_{\alpha\beta}$ where $\alpha\beta = wn, ws$, or ns:

$$T_w = T_s; \quad \underline{n}_{ws} \cdot \lambda_w \nabla T_w = \underline{n}_{ws} \cdot \lambda_s \nabla T_s \quad \text{at } A_{ws} \quad (3.113)$$

$$T_n = T_s; \quad \underline{n}_{ns} \cdot \lambda_n \nabla T_n = \underline{n}_{ns} \cdot \lambda_s \nabla T_s \quad \text{at } A_{ns} \quad (3.114)$$

$$T_w = T_n; \quad \underline{n}_{wn} \cdot (-\lambda_n \nabla T_n + \rho_n h_n (\underline{v}_n - \underline{v}_{wn}))$$
$$= \underline{n}_{wn} \cdot (-\lambda_w \nabla T_w + \rho_w h_w (\underline{v}_w - \underline{v}_{wn})) \quad \text{at } A_{wn}, \quad (3.115)$$

where $\underline{n}_{\alpha\beta}$ is the unit vector normal to the $\alpha\beta$-interface pointing out of α into β.

Upscaling (here volume averaging) of the pore-scale energy balance equations as given in Duval et al. [2004] then yields

$$\underbrace{\frac{\partial (\phi \rho_w S_w \langle h_w \rangle^w)}{\partial t}}_{1} + \underbrace{\nabla \cdot (\phi \rho_w S_w \langle h_w \rangle^w \langle \underline{v}_w \rangle^w)}_{2} + \underbrace{\nabla \cdot (\rho_w \langle h'_w \underline{v}'_w \rangle)}_{3}$$

$$+ \underbrace{\frac{1}{V} \int_{A_{wn}} \underline{n}_{wn} \rho_w h_w \cdot (\underline{v}_w - \underline{v}_{wn}) \, dA}_{4}$$

$$= \underbrace{\nabla \cdot \langle \lambda_w \nabla T_w \rangle}_{5} + \underbrace{\frac{1}{V} \int_{A_{wn}} \underline{n}_{wn} \lambda_w \cdot \nabla T_w \, dA}_{6} + \underbrace{\frac{1}{V} \int_{A_{ws}} \underline{n}_{ws} \lambda_w \cdot \nabla T_w \, dA}_{7} \quad (3.116)$$

$$\frac{\partial (\phi \rho_n S_n \langle h_n \rangle^n)}{\partial t} + \nabla \cdot (\phi \rho_n S_n \langle h_n \rangle^n \langle \underline{v}^n \rangle_n) + \nabla \cdot (\rho_n \langle h'_n \underline{v}'_n \rangle)$$

$$+ \frac{1}{V} \int_{A_{wn}} \underline{n}_{wn} \rho_n h_n \cdot (\underline{v}_n - \underline{v}_{wn}) \, dA$$

$$= \nabla \cdot \langle \lambda_n \nabla T_n \rangle + \frac{1}{V} \int_{A_{wn}} \underline{n}_{wn} \lambda_n \cdot \nabla T_n \, dA + \frac{1}{V} \int_{A_{ns}} \underline{n}_{ns} \lambda_n \cdot \nabla T_n \, dA \quad (3.117)$$

$$\rho_s C_{p,s} \frac{\partial (1-\phi) \langle T_s \rangle_s}{\partial t}$$

$$= \nabla \cdot \langle \lambda_s \nabla T_s \rangle + \frac{1}{V} \int_{A_{ws}} \underline{n}_{ws} \lambda_s \cdot \nabla T_s \, dA + \frac{1}{V} \int_{A_{ns}} \underline{n}_{ns} \lambda_s \cdot \nabla T_s \, dA. \quad (3.118)$$

The brackets $\langle \cdot \rangle$ indicate a macro-scale (volume-averaged) quantity while the primes denote a deviation term (the difference between pore-scale and volume-averaged quantity). The variable $C_{p,s}$ denotes the specific heat capacity of the solid phase at isobaric conditions. The storage term (terms of type 1) and the advective energy transport term (type 2) contain only averaged, macro-scale quantities. All the other terms still contain pore-scale parameters, either original pore-scale parameters or deviation terms, and need further discussion.

While Duval et al. [2004] closed their system of equations by solving a pore-scale problem, Niessner and Hassanizadeh [2009c] proceeded by searching for a macro-scale closure as the pore-scale situation is generally unknown.

Terms of type 3 and 5 are often combined as both are of diffusive type. Their combination is approximated as macro-scale thermal dispersion

$$\nabla \cdot (\rho_w \langle h'_w \underline{v}'_w \rangle) - \nabla \cdot \langle \lambda_w \nabla T_w \rangle \approx -\nabla \cdot \left(\underline{\underline{D}}^{th}_w \nabla \langle T_w \rangle^w \right) \tag{3.119}$$

with the thermal dispersion tensor given by

$$\underline{\underline{D}}^{th}_w = \lambda_w \underline{\underline{I}} + \alpha_T \rho_w C_{p,w} |\underline{v}_w| \underline{\underline{I}} + (\alpha_L - \alpha_T) \rho_w C_{p,w} \frac{\underline{v}_w \otimes \underline{v}_w}{|\underline{v}_w|} \tag{3.120}$$

containing longitudinal (α_L) and transversal (α_T) dispersivities. The term $\underline{\underline{I}}$ denotes the unit tensor while $C_{p,\alpha}$ is the specific heat capacity of phase α at isobaric conditions.

Terms of type 4 can be further transformed in the following way:

$$\int_{A_{wn}} \underline{n}_{wn} \rho_w h_w \cdot (\underline{v}_w - \underline{v}_{wn}) \, dA = \int_{A_{wn}} \underline{n}_{wn} \cdot \rho_w C_{p,w} (T^{sat} - \langle T_w \rangle^w)$$
$$\cdot (\underline{v}_w - \underline{v}_{wn}) \, dA \tag{3.121}$$

where T^{sat} is the saturation temperature. The right hand side may be written as (given the fact that $C_{p,w}$, T^{sat}, $\langle T_w \rangle^w$ are constant within an REV):

$$\int_{A_{wn}} \underline{n}_{wn} \rho_w h_w \cdot (\underline{v}_w - \underline{v}_{wn}) \, dA = C_{p,w} (T^{sat} - \langle T_w \rangle^w)$$
$$\int_{A_{wn}} \underline{n}_{wn} \cdot \rho_w (\underline{v}_w - \underline{v}_{wn}) \, dA \tag{3.122}$$
$$\approx \frac{1}{2} C_{p,w} (T^{sat} - \langle T_w \rangle^w) \Gamma_{wn} E_{wn}, \tag{3.123}$$

The integral over A_{wn} in Eq. (3.122) is actually the mass exchange between the wetting phase and the wn-interface. We have assumed that this term accounts for half of the wn-interface generation. Thus, it is replaced by $\frac{1}{2} E_{wn}$ where E_{wn} denotes the rate of production of wn-interfaces. In Niessner and Hassanizadeh [2008] it was conjectured that this E_{wn} can be a function of the time rate of change of wetting-phase saturation, $\frac{\partial S_w}{\partial t}$, and of capillary pressure p_c and saturation S_w itself through a function $e_{wn}(S_w, p_c)$. The value of e_{wn} could be obtained along the bounding $p_c(S_w)$ functions and interpolated in between. The parameter Γ_{wn} is the mass density of the interface. More information about the production functions E_{wn} and e_{wn} will be given in Sec. 5.1.1.2.

Terms 6 and 7 can be treated analogously to the diffusive part of mass transfer, e.g. terms of type 6:

$$\int_{A_{wn}} \underline{n}_{wn} \lambda_w \cdot \nabla T_w \, dA = \int_{A_{wn}} \lambda_w \frac{\partial T_w}{\partial n} \, dA = \int_{A_{wn}} \lambda_w \frac{T|_{wn} - T|_w}{d_w^T} \, dA$$

$$= \frac{a_{wn} \lambda_w}{d_w^T} \left(T^{sat} - \langle T_w \rangle^w \right), \quad (3.124)$$

and type 7

$$\int_{A_{ws}} \underline{n}_{ws} \lambda_w \cdot \nabla T_w \, dA = \int_{A_{ws}} \lambda_{ws} \frac{T|_s - T|_w}{d_{ws}^T} \, dA = \frac{a_{ws} \lambda_{ws}}{d_{ws}^T} \left(\langle T_s \rangle^s - \langle T_w \rangle^w \right), \quad (3.125)$$

where d_α^T and $d_{\alpha\beta}^T$ are thermal diffusion lengths of phase α and the $\alpha\beta$-interface, respectively, and $\lambda_{\alpha\beta}$ is approximated as $\lambda_{\alpha\beta} = \frac{1}{2}(\lambda_\alpha + \lambda_\beta)$. As the mass of interfaces is small compared to the mass of the bulk phases, we consider neither T^{sat} nor $T_{\alpha\beta}$ to be relevant quantities. Instead, for the fluid–solid interfaces, we directly parameterize the difference between adjacent phase temperatures without taking the intermediate (interface) value into account.

This also leads to a simpler form of term 4 that was approximated by Eq. (3.123):

$$0.5 \cdot C_{p,w}(T^{sat} - \langle T_w \rangle^w) \Gamma_{wn} E_{wn} \approx 0.25 \cdot C_{p,wn}(\langle T_n \rangle^n - \langle T_w \rangle^w) \Gamma_{wn} E_{wn}, \quad (3.126)$$

where we will make the approximation $C_{p,wn} = \frac{1}{2}(C_{p,w} + C_{p,n})$. Based on these simplifications, a complete macro-scale set of balance equations, constitutive relationships, and closure conditions will be given in Sec. 5.5 and used for numerical modeling of two-phase flow including kinetic interphase mass and energy transfer.

As the macro-scale energy balance equations not only contain the specific fluid–fluid interfacial area a_{wn} as a parameter, but additionally the fluid–solid interfacial areas a_{ws} and a_{ns}, constitutive relationships have to be specified for these variables as well.

Constitutive relationships between fluid–fluid specific interfacial area, capillary pressure, and saturation have already been derived either from pore-scale network models (Reeves and Celia [1996], Held and Celia [2001], Joekar-Niasar et al. [2008])

or from experiments (Brusseau et al. [1997], Chen and Kibbey [2006], Culligan et al. [2004], Schaefer et al. [2000], Wildenschild et al. [2002], Chen et al. [2007]). Relationships for fluid–solid specific interfacial areas could in principle be obtained exactly the same way. In the following, we present very recent results by Ahrenholz et al. [2009] who obtain $a_{ws}(S_w, p_c)$ and $a_{ns}(S_w, p_c)$ relationships from Lattice-Boltzmann simulations. As data for fluid–solid interfacial areas are still extremely scarce, we propose an approximation of these areas as an alternative.

Specific solid–fluid interfacial area surfaces from Lattice-Boltzmann simulations In the following, it is assumed that the solid grain surfaces are perfectly wettable, i.e. they are always covered by the wetting phase. So, in principle, there are no ns-interfaces and the specific interfacial area for ws may be assumed to be equal to the specific surface area of the solid, a_s. When the medium is fully saturated with the wetting phase, then $a_{wn} = 0$. At the other end, when the medium is almost filled with the non-wetting phase, then the wetting phase forms a film on the solid grains. Then, we may write

$$a_{ws} = a_s = a_{wn}. \tag{3.127}$$

However, such a wn-interface is not really controlled by capillary forces, but by adhesion forces. Therefore, it is possible to treat the wetting-phase film as part of the solid surface. In this case, it may be considered to be a non-wetting–solid interface whose properties are modified by the presence of the wetting phase.

Ahrenholz et al. [2009] performed quasi-static Lattice-Boltzmann simulations for air–water flow in a porous medium whose structure was provided by a CT scan of a real porous medium. They simulated primary drainage as well as main imbibition and main drainage, see Fig. 3.9. For a large number of quasi-static states, they determined capillary pressure and saturation. Fluid–fluid as well as fluid–solid interfacial areas could be obtained from a triangulation of the surfaces for the same states. Ahrenholz et al. [2009] used this data in order to obtain $a_{ws}(S_w, p_c)$ and $a_{ns}(S_w, p_c)$ surfaces, see Fig.s 3.10 and 3.11.

Approximation of specific solid–fluid interfacial area surfaces Often, plots of wn-interfaces show graphs similar to the ones shown in Fig. 3.12. Here, the curve a_{wn} denotes the interfacial area formed between the bulk wetting phase and the bulk

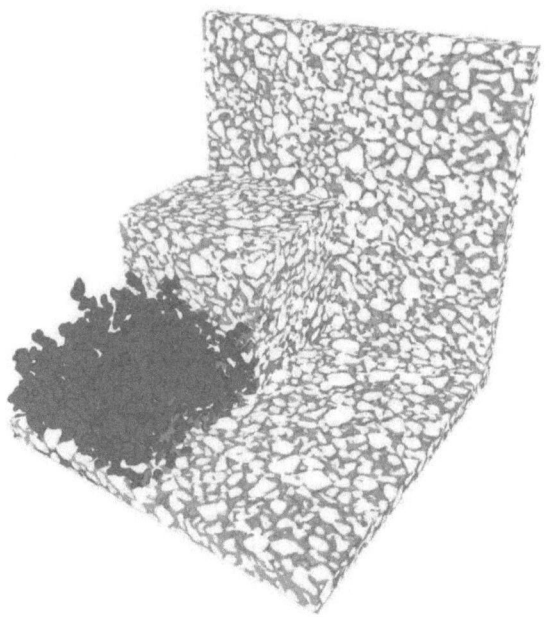

Figure 3.9: Lattice-Boltzmann simulation of a drainage process in a porous medium (taken from Ahrenholz et al. [2009]).

non-wetting phase. In addition to this, we have an interface between a wetting film and the non-wetting phase. The sum of the wn-interfaces is the monotonic curve denoted by $a_{wn,tot}$. The difference $a_{wn,tot} - a_{wn}$ accounts for wetting-film interfaces with the non-wetting phase. As explained above, we choose to consider these as ns-interfaces. Thus, the equality

$$a_{ns} = a_{wn,tot} - a_{wn} \tag{3.128}$$

has to hold. This implies that

$$a_{ws} = a_s - a_{ns} = a_s - a_{wn,tot} + a_{wn}. \tag{3.129}$$

We first need to calculate $a_{wn,tot}$ in order to obtain a_{ws}. Therefore, we construct the tangent to the $a_{wn}(S_w)|_{p_c^0}$ function for a given capillary pressure value of p_c^0 that

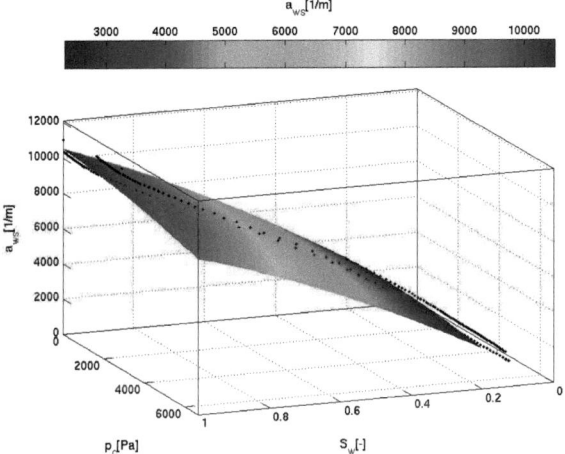

Figure 3.10: Relationship $a_{ws}(S_w, p_c)$ as obtained by Ahrenholz et al. [2009] from Lattice-Boltzmann simulations of air–water flow in a natural porous medium.

goes through the point $(0, a_s)$ in order to estimate the function $a_{wn,tot}(S_w)|_{p_c^0}$ for all saturation values smaller than the tangent point saturation S_w^*, see Fig. 3.12. For all saturations larger than S_w^*, we approximate $a_{wn,tot} = a_{wn}$, $a_{ws} = a_s$, and $a_{ns} = 0$ as indicated by Fig. 3.12. This procedure allows us to estimate fluid–solid interfaces using the following two assumptions:

1. The maximum total specific wn-interfacial area is approximately equal to the total specific interfacial area of the solid phase, $\max(a_{wn,tot}) \approx a_s$

2. The total interfacial area function is linear as soon as wn-interfaces in wetting-phase films on the solid surface are created.

Starting from a bi-quadratic formula for a_{wn} (see Joekar-Niasar et al. [2008])

$$a_{wn}(S_w, p_c^0) := a_{wn}(S_w)|_{p_c^0} = \underbrace{(a_{00} + a_{01}p_c^0 + a_{02}(p_c^0)^2)}_{b_0} + \underbrace{(a_{10} + a_{11}p_c^0)}_{b_1}S_w + \underbrace{a_{20}}_{b_2}S_w^2 \quad (3.130)$$

it follows that the tangent to this function at a saturation S_w^*, passing through the point $(0, a_s)$ must fulfill

$$a_{wn,tot}(S_w)|_{p_c^0} = (b_1 + 2b_2 S_w^*)S_w + a_s. \quad (3.131)$$

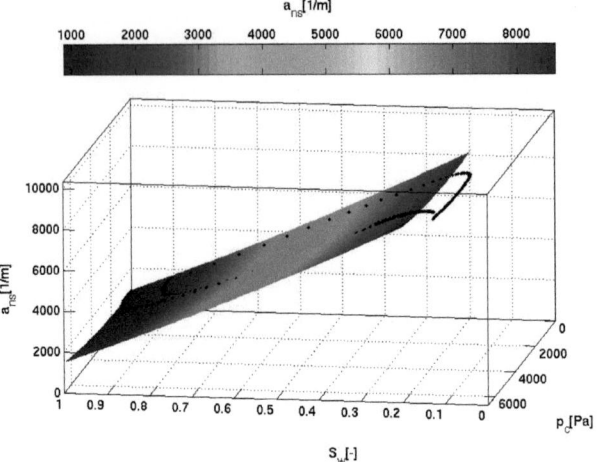

Figure 3.11: Relationship $a_{ns}(S_w, p_c)$ as obtained by Ahrenholz et al. [2009] from Lattice-Boltzmann simulations of air-water flow in a natural porous medium.

Using the condition that this tangent must also go through a point $(S_w^*, b_0 + b_1 S_w^* + b_2 S_w^{*2})$ on the $a_{wn,tot}(S_w)|_{p_c^0}$ curve, the value of S_w^* can be obtained from

$$b_0 + b_1 S_w^* + b_2 S_w^{*2} = b_1 S_w^* + 2b_2 S_w^{*2} + a_s, \quad (3.132)$$

which gives $S_w^* = \sqrt{\frac{b_0 - a_s}{b_2}}$. We then have all the information to actually obtain the desired relationships:

If $(S_w < S_w^*)$

$$a_{wn,tot}(S_w)|_{p_c^0} = (b_1 + 2\sqrt{b_2(b_0 - a_s)})S_w + a_s \quad (3.133)$$

$$a_{ws}(S_w, p_c) = -(b_1 + 2\sqrt{b_2(b_0 - a_s)})S_w + a_{wn}(S_w, p_c) \quad (3.134)$$

$$a_{ns}(S_w, p_c) = a_s + (b_1 + 2\sqrt{b_2(b_0 - a_s)})S_w - a_{wn}(S_w, p_c) \quad (3.135)$$

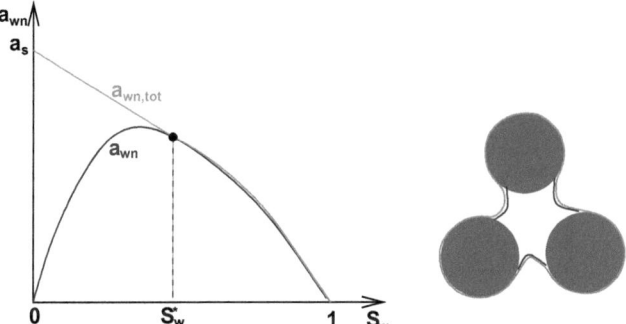

Figure 3.12: Estimation of a_{ws} and a_{ns} at a given p_c from an $a_{wn}(S_w, p_c)$ function: the inner lines in the right-hand picture show interfaces formed between the bulk wetting phase and the non-wetting phase; the outer line shows all wn-interfaces including interfaces between a wetting film and the non-wetting phase.

else

$$a_{wn,tot}(S_w)\big|_{p_c^0} = a_{wn}(S_w, p_c^0) \qquad (3.136)$$

$$a_{ws}(S_w, p_c) = a_s \qquad (3.137)$$

$$a_{ns}(S_w, p_c) = 0. \qquad (3.138)$$

In this section, we have presented a full macro-scale model based on the alternative approach which is able to account for kinetic interphase mass and energy transfer in a natural way, based on phase-interfacial areas. Thus, unlike in the classical approach, local chemical and thermal non-equilibrium can be handled.

3.6 Summary

In this chapter, an overview of alternatives to the classical model for two-phase flow in porous media was given. Then, a specific approach developed by Hassanizadeh & Gray was chosen which is based on rational thermodynamics and is thus, more physically founded than the classical two-phase model. A main principle of this approach is to formulate balance equations not only for bulk phases, but also for phase-interfaces and for common lines. Mass, momentum, energy, and entropy balance equations for phases and phase-interfaces were upscaled from the pore scale to the macro scale. In order to close the resulting system of macro-scale partial differential equations, constitutive equations were derived based on the entropy inequality. It was discussed in detail how the alternative model can be used to describe kinetic interphase mass and energy transfer.

4 Numerical modeling

The partial differential equations describing multi-phase and multi-phase multi-component processes that were discussed in Chap.s 2 and 3 do not have an analytical solution. Therefore, one can discretize the equations and solve them numerically at certain points in space corresponding to an underlying grid. A similar problem applies to the temporal resolution: it is impossible to give a solution at each point in time. Instead, the partial differential equations are also discretized in time meaning that they are solved at a number of discrete points in time.

Ideally, one desires to have a numerical scheme that is accurate and computationally cheap. More specifically, modelers usually have the following "wishes" for their numerical schemes:

- is locally mass conservative
- is higher order convergent
- can handle both elliptic and hyperbolic problems
- can take large accurate time steps
- can handle general unstructured non-matching grids.

So far, there is no numerical scheme that fulfills all these wishes. Therefore, in practice, modelers always need to find a compromise and choose a numerical scheme that ideally meets the needs of their specific model problem. In the following (Sec. 4.1), a short characterization of the most common discretization schemes is given. For each choice, a number of different options is available that adjust the scheme to the specific needs. It is not the purpose of this work to discuss all these different flavors as the focus of this work is on thermodynamically consistent modeling along with multi-scale multi-physics approaches and not on the development of numerical methods. Therefore, a short overview of classical methods is given and possibilities of the mathematical formulation of the governing equations (fully coupled versus fractional flow formulation) are discussed. Different time discretization possibilities are given in Sec. 4.2. In Sec. 4.3, the space discretization method used in this work is presented.

4.1 Overview of space discretization methods and mathematical formulations

4.1.1 Discretization methods

In principle, one can distinguish Eulerian and Langrangian methods. **Lagrangian methods** follow a particle along its path. This means they use a moving frame of reference as the particles move from their initial location. It is said that an observer of a Lagrangian model follows along with the flow.

Eulerian methods, contrarily, use a fixed frame of reference and balance conserved properties (mass, momentum, energy, and entropy) with respect to this control volume. Most discretization methods used in fluid mechanics are based on an Eulerian approach.

The following classical discretization methods can be distinguished. Unless otherwise specified, they are based on the Eulerian approach.

- **finite differences.** Finite difference schemes directly discretize the partial derivatives occuring in the partial differential equations by finite differences, see e.g. Ames [1977]. The derivation is based on a Taylor expansion where usually, only zeroth and first order terms are kept. The inclusion of higher order terms is also possible. Main differences between the methods result from the fact whether forward, backward, or central differences are applied. This choice also determines the order of the scheme. Finite difference methods are easy to understand and implement. A major drawback of these methods is that they are not mass conservative (neither locally nor globally). This is especially crucial for solving transport problems where e.g. the mass of a contaminant is a quantity needed to be known in high accuracy and any unphysical decay or production of contaminants should be avoided by all means.

- **finite volumes.** Finite volume methods are based on the Eulerian approach in its most intuitive form. Here, conserved properties (mass, momentum, and energy) are balanced over control volumes, see Hirsch [1988]. The main advantage of this procedure is that local mass conservation is automatically guaranteed. It is, however, not straight forward to combine finite volume methods with unstructured grids. For this extension, the method clearly needs a special adaptation.

- **finite elements.** The idea of finite element methods is to integrate the residal error over the whole model domain, such that it becomes zero on average, see e.g. Aziz and Settari [1979]. This means that the scheme is globally, but not necessarily locally, mass conservative. A major advantage is that it can be easily applied to unstructured grids. A large number of different options is available that partly make the scheme better suited for elliptic or hyperbolic problems, allow for easy higher order extensions, or make the scheme locally mass conservative. Standard Galerkin, Petrov-Galerkin, fully upwinding, mixed hybrid, or discontinuous Galerkin finite element methods represent such tailored adaptations.

- **method of characteristics.** The method of characteristics can be applied to first order partial differential equations and reduces them to a family of ordinary differential equations along which the solution can be integrated from some initial data, see e.g. Russell [1985]. The method of characteristics discovers curves (called characteristics) along which the partial differential equation becomes an ordinary differential equation. Once the ordinary differential equation is found, it can be solved along the characteristics and transformed back into a solution of the original partial differential equation. The method is computationally cheap and easy, but its range of application is limited.

- **Eulerian–Lagrangian localized adjoint method (ELLAM).** ELLAM methods represent an approach that allows to capture the fact that transport problems are generally Lagrangian in nature (Binning and Celia [1996], Ewing and Wang [1994]). The Eulerian framework, however, helps to still gurantee local mass conservation. ELLAM schemes are computationally expensive, but allow for highly accurate solutions.

- **streamline methods.** Streamline methods are well-suited for hyperbolic problems. The strategy is to trace streamlines based on the velocity field and to solve one-dimensional transport problems along these streamlines, see e.g. Pollock [1988]. The method is extremly fast and accurate for losely coupled flow and transport problems. However, for elliptic and parabolic problems or if the coupling between flow and transport is significant, streamline methods are not well-suited.

4.1.2 Fully coupled versus fractional flow formulation

In this section, two basically different mathematical formulations of the two-phase flow equations are discussed, the fully coupled and the fractional flow formulation. Although they represent mathematical formulations and not numerical methods themselves, the chosen mathematical formulation represents the starting point for any discretization. It is therefore essential to discuss the formulation possiblities before discussing discretization methods. As the interfacial-area-based approach is so far only used in the framework of the fully coupled formulation (the reasons therefore will become obvious in the course of this chapter), the different formulations are discusseed in the frame of the classical approach.

Considering the example of a multi-phase, say, an m-phase system in the frame of the classical approach, one has the possibility either to solve the m balance equations for the m-phase system directly in the form of Eq. (2.35). In this case, there are m balance equations that are all coupled to one another. This approach is called the *fully coupled* formulation and is further studied in Sec. 4.1.2.1. An analogous formulation can be constructed for multi-phase multi-component systems.

The other possibility for multi-phase systems in the frame of the classical approach is to transform the system of m balance equations into one equation for pressure, the *pressure equation*, and $(m-1)$ equations for selected saturations, the *saturation equations*. This approach is the so-called *fractional flow formulation*. These $(1 + (m-1))$ equations are usually solved sequentially. The easiest and most common approach is the IMPES scheme (IMplicit pressure, Explicit Saturation), where the pressure equation is first solved implicitly to yield the velocity field which then enters into the $(m-1)$ saturation equations. Like the fully coupled formulation, the fractional flow formulation can also be applied to multi-phase multi-component systems. An extension to the interfacial-area-based approach is also possible in principle, but its usefulness in terms of physics may be questioned. The reason is that the fractional flow formulation is advantageous if capillary pressure effects are small as then, pressure and saturation equation(s) are not strongly coupled and a sequential solution makes sense. However, the interfacial-area-based approach is especially advantageous if capillary pressure effects (including hysteresis) are strong and thus, the presence of interfacial areas in the equations can be exploited.

An overview of different formulations for multi-phase multi-component systems with varying numbers of phases and components is given in Peaceman [1977]. Binning and Celia [1999] make a detailed study on the advantages and disadvantages of the fractional flow formulation for one-dimensional systems. The fractional flow formulation for both multi-phase and for multi-phase multi-component systems is presented in Sec. 4.1.2.2.

4.1.2.1 Fully coupled formulation

In this section, the fully coupled formulation for both multi-phase and for multi-phase multi-component models is reviewed in the frame of the classical approach.

Multi-phase systems.

First, general m-phase flow is considered. Inserting the extended Darcy's law,

$$\underline{v}_\alpha = -\underline{\underline{K}} \frac{k_{r\alpha}(S_\alpha, \ldots)}{\mu_\alpha} \left(\nabla p_\alpha - \rho_\alpha \underline{g} \right) \tag{4.1}$$

into the mass balance given by Eq. (2.35) gives m equations, one for each of the m phases:

$$\frac{\partial (\phi \rho_\alpha S_\alpha)}{\partial t} - \nabla \cdot \left(\rho_\alpha \frac{k_{r\alpha}}{\mu_\alpha} \underline{\underline{K}} (\nabla p_\alpha - \rho_\alpha \underline{g}) \right) - \rho_\alpha q_\alpha = 0. \tag{4.2}$$

The unknown quantities are the pressures of the m phases, $p_1, \ldots, p_m$, and the phase saturations, $S_1, \ldots, S_m$. That means, only m equations for $2m$ unknowns can be formulated. The common strategy to tackle this problem is to choose m primary variables among the $2m$ unknowns and to use m closure relations to substitute the m secondary unknowns. In the following considerations, it is assumed that the pressures, as well as the saturations are ordered according to their wettability, i.e. fluid 1 is the most wetting fluid, and fluid m the least wetting fluid. Then, these m closure relations are postulated to be given by

$$\sum_{i=1}^{m} S_i = 1 \qquad (4.3)$$

$$p_{c12} = p_2 - p_1 = p_c(S_1) \qquad (4.4)$$

$$p_{c23} = p_3 - p_2 = p_c(S_2) \qquad (4.5)$$

$$\vdots = \vdots$$

$$p_{c(m-1)m} = p_m - p_{m-1} = p_c(S_{m-1}), \qquad (4.6)$$

i.e. the m closure relations consist of the condition that the sum of the saturations has to be equal to 1, and $(m - 1)$ capillary pressure–saturation relationships that relate the phase pressures to saturations.

Note, that for the fully coupled approach, the m differential equations are all of the same structure. Mathematically, they represent a parabolic nonlinear partial differential equation system.

Multi-phase multi-component systems.

Similar considerations can be made for multi-phase multi-component flow and transport. Here, a system with m phases and m components is considered. Inserting the extended Darcy's law given in Eq. (4.1) into the mass balance equation for multi-phase multi-component systems of Eq. (2.41) results in

$$\frac{\partial C^\kappa}{\partial t} - \sum_\alpha \nabla \cdot \left(C_\alpha^\kappa \frac{k_{r\alpha}}{\mu_\alpha} \underline{\underline{K}} (\nabla p_\alpha - \rho_\alpha \underline{g}) - \underline{\underline{D}}_{pm}^\kappa \nabla C_\alpha^\kappa \right) - q^\kappa = 0. \qquad (4.7)$$

In this case, there are m governing equations for $m^2 + m$ unknowns (m pressures, and m^2 concentrations C_α^κ. To close the equation system, the same m conditions as for multi-phase flow are usually postulated

$$\sum_{i=1}^{m} S_i = 1 \tag{4.8}$$

$$p_{c12} = p_2 - p_1 = p_c(S_1) \tag{4.9}$$

$$p_{c23} = p_3 - p_2 = p_c(S_2) \tag{4.10}$$

$$\vdots = \vdots \tag{4.11}$$

$$p_{c(m-1)m} = p_m - p_{m-1} = p_c(S_{m-1}), \tag{4.12}$$

and again, m independent primary variables are chosen. With the definition of total concentration,

$$C^\kappa = \phi \sum_\alpha (\rho_\alpha S_\alpha X_\alpha^\kappa), \tag{4.13}$$

m equations are given, but at the same time, m^2 new unknowns X_α^κ are introduced. However, it is known from Eq. (2.4) that

$$C_\alpha^\kappa = \rho_\alpha \cdot X_\alpha^\kappa. \tag{4.14}$$

Eq. (2.2) gives m conditions for the mass fractions X_α^κ:

$$\sum_\kappa X_\alpha^\kappa = 1, \tag{4.15}$$

and then, one has to formulate equations of state to relate the mass fractions X_α^κ or the mole fractions x_α^κ to the primary variables. This is usually done by assuming local chemical equilibrium and using Henry's and Raoult's laws, see Sec. 2.4.4.

The key point for the fully coupled multi-phase multi-component system is again, that the partial differential equation system is solved in the form of Eq. (4.7), i.e. as a parabolic, nonlinear, and coupled system of equations, that are all of the same type.

4.1.2.2 Fractional flow formulation

This section deals with the fractional flow formulation for both multi-phase and for multi-phase multi-component systems in the frame of the classical approach, see Peaceman [1977].

Multi-phase systems.

It is not as straightforward as for the fully coupled formulation to derive a general representation of the fractional flow formulation for an m-phase system. Therefore, as an example, a two-phase systems is discussed, i.e., a wetting phase w and a non-wetting phase n are focussed on. Details on the this formulation, also for three-phase flow, can be found in the work of Chen and Ewing [1997] and Huber [2000]. First, the two-phase flow equation system in the form of Eq. (2.35) is considered,

$$\frac{\partial(\phi \rho_\alpha S_\alpha)}{\partial t} + \nabla \cdot (\rho_\alpha \underline{v}_\alpha) - \rho_\alpha q_\alpha = 0, \tag{4.16}$$

with the closure relations

$$S_w + S_n = 1, \text{ and} \tag{4.17}$$

$$p_c = p_n - p_w = p_c(S_w), \tag{4.18}$$

and the extended Darcy velocity

$$\underline{v}_\alpha = -\frac{k_{r\alpha}}{\mu_\alpha} \underline{\underline{K}} (\nabla p_\alpha - \rho_\alpha \underline{g}). \tag{4.19}$$

Adding up Eq.s (4.16) for $\alpha = w$ and $\alpha = n$, one is left with

$$\nabla \cdot \sum_\alpha \underline{v}_\alpha = -\frac{\partial \phi}{\partial t} - \sum_\alpha \frac{1}{\rho_\alpha} \left(\phi S_\alpha \frac{\partial \rho_\alpha}{\partial t} + \underline{v}_\alpha \nabla \rho_\alpha - \rho_\alpha q_\alpha \right). \tag{4.20}$$

For the fractional flow formulation, two new variables are introduced, the *total velocity* $\underline{v}$, and the *global pressure* $\tilde{p}$. These new variables are "artificial" variables which are only defined for the fractional flow formulation, but do not represent measurable physical quantities. The sum of the phase velocities is denoted as the total velocity $\underline{v}$

$$\underline{v} = \sum_\alpha \underline{v}_\alpha. \tag{4.21}$$

The global pressure is defined in a way, that a relationship similar to Darcy's law can be established between global pressure and total velocity,

$$\underline{v} = -\lambda \underline{\underline{K}} \cdot (\nabla \tilde{p} - \underline{G}), \tag{4.22}$$

where λ is the total mobility defined as

$$\lambda = \sum_\alpha \lambda_\alpha. \qquad (4.23)$$

The phase mobilities λ_α are calculated as

$$\lambda_\alpha = \frac{k_{r\alpha}}{\mu_\alpha}, \qquad (4.24)$$

and $\underline{G}$ is the gravity term defined by

$$\underline{G} = \sum_\alpha \rho_\alpha f_\alpha \underline{g}. \qquad (4.25)$$

The variables f_α denote the fractional flow function of phase α which gave the name to this formulation. It is defined as

$$f_\alpha = \frac{\lambda_\alpha}{\lambda}. \qquad (4.26)$$

Following Chen and Ewing [1997] and Chavent and Jaffré [1986], the global pressure is given by

$$\tilde{p} = p_n - \int_{S_c}^{S_w} f_w \frac{dp_c(\xi)}{dS_w} d\xi, \qquad (4.27)$$

where S_c is defined such that $p_c(S_c) = 0$. The integral in Eq. (4.27) has to be evaluated numerically, which means, that it either lacks accuracy or—when accurately computing the integral—that the calculation is time consuming. Having this information, already one of the two equations of a fractional flow formulation for two-phase flow is defined, the *pressure equation*. The total set of equations defining the pressure equation is

$$\nabla \cdot \underline{v} = -\frac{\partial \phi}{\partial t} - \sum_\alpha \frac{1}{\rho_\alpha} \left(\phi S_\alpha \frac{\partial \rho_\alpha}{\partial t} + \underline{v}_\alpha \nabla \rho_\alpha - \rho_\alpha q_\alpha \right) \qquad (4.28)$$

$$\underline{v} = -\lambda \underline{K} \cdot (\nabla \tilde{p} - \underline{G}), \text{ and} \qquad (4.29)$$

$$\tilde{p} = p_n - \int_{S_c}^{S_w} f_w \frac{dp_c(\xi)}{dS_w} d\xi. \qquad (4.30)$$

The other equation to be defined is the *saturation equation*. One can choose to formulate the equation either for the wetting phase saturation S_w or for the non-wetting

phase saturation S_n. Here, the wetting phase saturation S_w is chosen.
Considering the definitions of the phase velocities

$$\underline{v}_w = -\lambda_w \underline{\underline{K}} \left(\nabla p_w - \rho_w \underline{g} \right) \tag{4.31}$$

$$\underline{v}_n = -\lambda_n \underline{\underline{K}} \left(\nabla p_n - \rho_n \underline{g} \right) = -\lambda_n \underline{\underline{K}} \left(\nabla p_w + \nabla p_c - \rho_n \underline{g} \right), \tag{4.32}$$

both equations can be resolved for $\underline{\underline{K}} \nabla p_w$ and equated. With the condition $\underline{v} = \underline{v}_w + \underline{v}_n$, one gets after resolving for $\underline{v}_w$

$$\underline{v}_w = f_w \underline{v} + \lambda_n f_w \left(\rho_w - \rho_n \right) \underline{\underline{K}} \underline{g} + \lambda_n f_w \underline{\underline{K}} \nabla p_c. \tag{4.33}$$

Inserting this expression into Eq. (4.16) with $\alpha = w$, one obtains the final saturation equation

$$\underbrace{\frac{\partial (\phi \rho_w S_w)}{\partial t}}_{1} + \underbrace{\nabla \cdot (\rho_w f_w \underline{v})}_{2} + \underbrace{\nabla \cdot (\rho_w \lambda_n f_w (\rho_w - \rho_n) \underline{\underline{K}} \underline{g})}_{3}$$

$$+ \underbrace{\nabla \cdot (\rho_w f_w \lambda_n \underline{\underline{K}} \nabla p_c)}_{4} - \underbrace{\rho_w q_w}_{5} = 0. \tag{4.34}$$

In this form, one can clearly identify the mathematical character of the different terms. The term 1 is the accumulation term, while the terms 2 and 3 have advective character, and 4 is of diffusive nature while term 5 is a sink / source term.

Multi-phase multi-component systems.

First, the differential equations for multi-phase multi-component flow and transport are considered, given by Eq. (2.41),

$$\frac{\partial C^\kappa}{\partial t} + \sum_\alpha \nabla \cdot (C_\alpha^\kappa \underline{v}_\alpha + \underline{\underline{D}}_{pm}^\kappa \nabla C_\alpha^\kappa) - q^\kappa = 0. \tag{4.35}$$

The pressure equation for a multi-phase multi-component model is identical to the pressure equation of a multi-phase model

$$\nabla \cdot \underline{v} = -\frac{\partial \phi}{\partial t} - \sum_\alpha \frac{1}{\rho_\alpha} \left(\phi S_\alpha \frac{\partial \rho_\alpha}{\partial t} + \underline{v}_\alpha \nabla \rho_\alpha - \rho_\alpha q_\alpha \right) \tag{4.36}$$

$$\underline{v} = -\lambda \underline{\underline{K}} \cdot (\nabla \tilde{p} - \underline{G}). \tag{4.37}$$

The concentration equations can be derived analogously to multi-phase flow. Following the derivations in Huber [2000], the resulting equations are

$$\frac{\partial C^\kappa}{\partial t} + \nabla \cdot \left[\sum_\alpha C_\alpha^\kappa f_\alpha \underline{v}\right] - \nabla \cdot \left[\sum_\alpha C_\alpha^\kappa (-1)^{\delta_{n\alpha}} \lambda_n f_w (\rho_w - \rho_n) \underline{\underline{K}} g\right]$$

$$- \nabla \cdot \left[\sum_\alpha C_\alpha^\kappa (-1)^{\delta_{n\alpha}} \lambda_n f_w \underline{\underline{K}} \nabla p_c\right] - \nabla \cdot \left[\phi \sum_\alpha S_\alpha \underline{\underline{D}}_{pm}^\kappa \cdot C_\alpha^\kappa\right] - q^\kappa = 0, \quad (4.38)$$

where $\delta_{n\alpha}$ is the Kronecker delta which is equal to 1 if $n = \alpha$, else it is zero.

4.2 Time discretization

With respect to time discretization, a principle distinction between explicit and implicit methods can be made.

Explicit methods calculate the state of a system at time t^{n+1} directly from the state of the system at the current time t^n, while **implicit methods** find a solution by solving an equation involving both t^n and t^{n+1}. It is clear that implicit methods require an extra computation step as a coupled system of equations needs to solved at each time step and they are much harder to implement. Still, implicit methods are most often used for practical applications because many problems arising in real life are stiff, for which the use of an explicit method requires impractically small time steps to keep the problem stable. For such problems, to achieve a given accuracy, it takes much less computational time to use an implicit method with larger time steps, even taking into account that one needs to solve an equation system at each time step. This means that the choice of an explicit or implicit method depends on the problem to be solved.

Time discretization methods for ordinary differential equations can be generalized as Runge–Kutta methods.

Explicit Runge–Kutta methods are given by

$$y_{n+1} = y_n + h \sum_{i=1}^{s} b_i k_i, \quad (4.39)$$

where

$$k_1 = f(t_n, y_n), \qquad (4.40)$$
$$k_2 = f(t_n + c_2 h, y_n + a_{21} h k_1), \qquad (4.41)$$
$$k_3 = f(t_n + c_3 h, y_n + a_{31} h k_1 + a_{32} h k_2), \qquad (4.42)$$
$$\vdots$$
$$k_s = f(t_n + c_s h, y_n + a_{s1} h k_1 + a_{s2} h k_2 + \cdots + a_{s,s-1} h k_{s-1}). \qquad (4.43)$$

To specify a particular method, one needs to provide the integer s (the number of stages) and the coefficients a_{ij} (for $1 \leq j < i \leq s$), b_i (for $i = 1, 2, \ldots, s$) and c_i (for $i = 2, 3, \ldots, s$).

The simplest example of an explicit Runge–Kutta method is the forward Euler method:

$$y_{n+1} = y_n + h f(t_n, y_n) \qquad (4.44)$$

Implicit Runge–Kutta methods, contrarily, contain more coefficients:

$$y_{n+1} = y_n + h \sum_{i=1}^{s} b_i k_i, \quad k_i = f\left(t_n + c_i h, y_n + h \sum_{j=1}^{s} a_{ij} k_j\right). \qquad (4.45)$$

Due to the fullness of the matrix a_{ij}, the evaluation of each k_i is now considerably involved and dependent on the specific function $f(t, y)$. Despite the difficulties, implicit methods are of great importance due to their high (possibly unconditional) stability, which is especially important in the solution of partial differential equations.

The simplest example of an implicit Runge–Kutta method is the backward Euler method:

$$y_{n+1} = y_n + h f(t_n + h, y_{n+1}) \qquad (4.46)$$

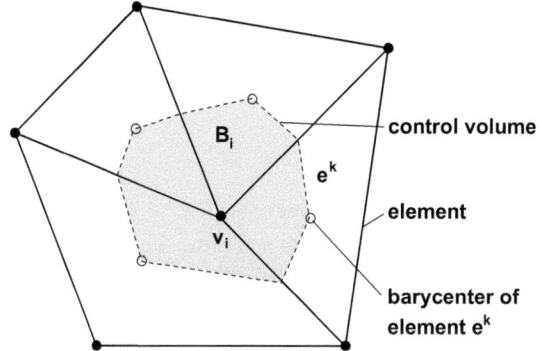

Figure 4.1: Construction of control volumes for a finite element grid.

4.3 Space discretization: vertex centered finite volume method

After these general considerations concerning space and time discretization, the method which is chosen for the numerical simulations in this work will be considered in further detail: the vertex-centered finite volume method (e.g. Bastian et al. [1997], Bastian and Helmig [1999]). In order prevent the appearance of non-physical sinks and sources, a locally conservative discretization is to be persued. But additionally, the solution of real-life problems is often challenging due to the complicated geometry involved. For these reasons, the vertex-centered finite volume scheme was developed. This discretization technique is not only locally conservative (like the finite volume method), but can also be applied to unstructured grids easily (like the finite element (FE) method).

For the vertex-centered finite volume method, two different grids are needed. Control volumes are constructed around the nodes of the initial FE mesh, which defines the elements. This construction is done by linking the barycenter of each element with the midpoints of the edges of this element, see Fig. 4.1.

We define a weighting function equal to 1 inside a control volume B_i, and equal to 0 outside this control volume:

$$W_i(x) = \begin{cases} 1 & \text{if } x \in B_i \\ 0 & \text{if } x \notin B_i. \end{cases} \quad (4.47)$$

The dimension of this space is equal to the number of nodes of the FE mesh, and the basis function (or shape function) associated with the ith vertex is N_i which is equal to 1 at node i of the FE mesh and equal to zero at all other nodes:

$$N_i(\text{node}_j) = \delta_{ij}, \qquad (4.48)$$

with δ_{ij} the Kronecker delta. The unknowns will be approximated in the space of piecewise polynomial functions which are first order (affine linear) on the triangles of the FE mesh and which are first order in each variable separately (affine bilinear) on each rectangle of the FE mesh.

Multiplying the two-phase flow equations (4.2) by the weighting function W_i and integrating over the whole domain G results in integrals over the control volumes:

$$\int_{B_i} \frac{\partial S_\alpha}{\partial t} \phi dB_i - \int_{B_i} \nabla \cdot (\lambda_\alpha K (\nabla p_\alpha - \rho_\alpha \mathbf{g})) \, dB_i - \int_{B_i} q_\alpha dB_i = 0, \qquad (4.49)$$

with the mobilities $\lambda_\alpha = \frac{k_{r\alpha}}{\mu_\alpha}$.

Applying the Green-Gaussian integral theorem with Γ_{B_i} as the boundary of B_i gives

$$\int_{B_i} \frac{\partial S_\alpha}{\partial t} \phi dB_i + \oint_{\Gamma_{B_i}} (\lambda_\alpha K (\nabla p_\alpha - \rho_\alpha \underline{g})) \cdot \underline{n} \, d\Gamma_{B_i} - \int_{B_i} q_\alpha dB_i = 0. \qquad (4.50)$$

Taking the shape functions into account yields the following form of the discretized two-phase flow equations:

$$\begin{aligned}
f_{\alpha i}(S_{ni}^{k+1}; S_{ni}^k; p_{wi}^{k+1}; p_{wm}^{k+1}) :=& \\
& - (-1)^{\delta_{\alpha w}} (S_{ni}^{k+1} - S_{ni}^k) \frac{\phi}{\Delta t} |B_i| \\
& - \sum_{l \in E_i} \sum_{j \in \eta_i} \lambda_{\alpha ij}^{FU_{el}} K^l \cdot \\
& \quad \cdot \sum_{m \in V} (p_{wm}^{k+1} \nabla N_m + \delta_{\alpha n} p_{cm}^{k+1} \nabla N_m \qquad (4.51) \\
& \quad - \rho_\alpha \underline{g} N_m) \cdot \underline{n}_{ij}^l \\
& - q_{\alpha i}^{k+1} |B_i| - m_{\alpha i} \\
=& \; 0,
\end{aligned}$$

with $\underline{n}_{ij}^l$ the outer normal vector. Here, η_i is the set of neighboring nodes whose control volumes share subcontrol volume edges/faces with B_i, E_i is the set of elements which have vertex v_i as a corner, and V is the set of all vertices of the FE mesh. $|B_i|$ represents the area (2D), respectively the volume (3D), of the control volume around node i, the indices k as well as $k+1$ denote the time step, and j represents a neighboring node of i. The integrals over the boundaries of B_i, Γ_{B_i}, are evaluated by using the midpoint rule; i.e., the integral over a segment of the boundary is calculated by multiplying the value at the midpoint of the control volume boundary segment by its length.

We set

$$\psi_{\alpha i}^{k+1} := p_{wi}^{k+1} + \delta_{\alpha n} p_{ci}^{k+1} - \rho_\alpha g z_i. \tag{4.52}$$

The sign of $\psi_{\alpha m} - \psi_{\alpha i}$ gives the direction of the flow of phase α across the interface between B_i and B_m. The term $m_{\alpha i}$ is the flow over $\partial B_i \cap \Gamma_{\alpha N}$ where $\Gamma_{\alpha N}$ is a Neumann type boundary for phase α. Using the fully upwind (FU) finite volume method for the mobilities results in

$$\lambda_{\alpha ij}^{FU_{e_l}} = \begin{cases} \lambda_{\alpha j} & \text{if } (\psi_{\alpha j} - \psi_{\alpha i}) \geq 0 \\ \lambda_{\alpha i} & \text{if } (\psi_{\alpha j} - \psi_{\alpha i}) < 0 \end{cases}, \tag{4.53}$$

which means that the mobility of the node with the higher potential is chosen.

The vertex-centered finite volume method is locally mass conservative, as exactly the same term occurs in the boundary integral of two neighboring control volumes.

The same discretization principle can be easily transferred both to the interfacial balance equation in case the interfacial-area-based approach is used and to multi-phase multi-component systems in case of compositional flow, simply by using analogous discretizations for analogous terms (storage, flux, and source term). It has to be noted, however, that for the flux term in the interfacial balance equation, central weighting is used instead of full upwinding. This is due to the fact that specific interfacial area is a non-monotonous function of saturation. This implies that in the absence of any external sinks or sources, an extremum in specific interfacial area can be due to physics while for saturation, these extrema can only be numerical artifacts.

4.4 Summary

In this chapter, a brief overview of classical spatial discretization methods was given and different mathematical formulations for both multi-phase and for multi-phase multi-component systems were discussed. Concerning time discretization, the difference between explicit and implicit methods was stressed and put into the context of general Runge-Kutta methods. Finally, the space discretization used for the simulation examples of this thesis (a vertex centered finite volume method) was explained in more detail.

5 Modeling examples

In this section, the alternative approach as introduced in Chap. 3 is applied to several modeling examples. First, the model is verified for isothermal immiscible two-phase flow by comparison to the classical model for a setup where hysteresis does not occur (Sec. 5.1). Then, in Sec. 5.2, the redistribution of two-fluid phases is studied and results are compared to a semi-analytical solution. Next, in Sec. 5.3, a setup for kinetic interphase mass transfer is studied and in Sec. 5.4, a similar setup including kinetic interphase energy transfer is investigated. In Sec. 5.5, an interface condition is introduced that guarantees a physically correct treatment of material interfaces. As a last example (Sec. 5.6), the application of the alternative model to a fracture–matrix system is studied. Additionally, a discussion on how relevant parameters for the alternative approach can be obtained either experimentally or numerically is given in Sec. 5.7. Finally, a chapter summary is provided in Sec. 5.8.

5.1 Isothermal immiscible two-phase flow

This section is based on the work of Niessner and Hassanizadeh [2008] and deals with the modeling of two-phase flow (isothermal and immiscible) in a homogeneous porous medium. Based on the work of Hassanizadeh and Gray [1979a,b, 1980, 1990], simplifications of the general mass and momentum balance equations were made in order to transform the general balance equations into a form which is handable by numerical models. In this spirit, the aim of Niessner and Hassanizadeh [2008] was to make the simplest possible extension of the classical two-phase flow model that still includes interfacial areas. Also, they proposed constitutive relationships for the relevant variables. In a simple numerical test example considering primary drainage under isothermal conditions for two immiscible phases, results are compared to results obtained using the classical model. The idea behind is to verify the interfacial-area-based model by comparison to the classical model as, without hysteresis, both models should give equal results.

Based on physical intuition and some common knowledge of typical two-phase flow systems, Niessner and Hassanizadeh [2008] propose the following simplifying assumptions of the general mass and momentum balance equations for phases and interfaces:

1. Porosity does not change with time ($\frac{\partial \phi}{\partial t} = 0$).

2. Phases are incompressible ($\rho_\alpha = $ const.).

3. Interfacial mass density $\Gamma_{\alpha\beta}$ is constant.

4. The effect of gravity on interfacial movement is neglected $\left(\Gamma_{wn} a_{wn} \underline{g} = \underline{0}\right)$.

5. The terms involving the derivative of Helmholtz free energies of phases and interfaces for saturation are neglected. This means, the only driving force for flow of phases is the hydraulic head gradient and for interfaces, the driving force is the gradient of specific interfacial area.

6. Mass exchange between phases and interfaces has negligible effect on mass balance of phases.

7. Cross-coupling terms in flow velocity among phases and interfaces are negligible.

8. The porous medium is assumed to be perfectly wettable to the wetting phase and, therefore, the role of solid–fluid interfaces (a_{ws} and a_{ns}) is neglected.

These simplifications allow us to neglect second-order effects such as medium deformation and fluid compressibility, keep the traditional two-phase flow equations, and introduce specific interfacial area into the model. After various transformations of the general mass and momentum balance equations given in Sec.s 3.2.2 and 3.2.3 and using the constitutive relationships obtained from exploiting the entropy inequality in Sec. 3.3, the following smaller set of equations consisting of continuity equations for the two phases and a balance of specific interfacial area for the fluid–

fluid interface are obtained:

$$\phi \frac{\partial S_w}{\partial t} + \nabla \cdot \underline{v}_w = Q_w, \quad (5.1)$$

$$\text{with} \quad \underline{v}_w = -\frac{S_w^2 \underline{\underline{K}}}{\mu_w} (\nabla p_w - \rho_w \underline{g})$$

$$\phi \frac{\partial S_n}{\partial t} + \nabla \cdot \underline{v}_n = Q_n, \quad (5.2)$$

$$\text{with} \quad \underline{v}_n = -\frac{S_n^2 \underline{\underline{K}}}{\mu_n} (\nabla p_n - \rho_n \underline{g})$$

$$\frac{\partial a_{wn}}{\partial t} + \nabla \cdot (a_{wn} \underline{v}_{wn}) = E_{wn}, \quad (5.3)$$

$$\text{with} \quad \underline{v}_{wn} = -\underline{\underline{K}}_{wn} \cdot \nabla a_{wn} \quad (5.4)$$

$$S_w + S_n = 1 \quad (5.5)$$

$$p_n - p_w = p_c \quad (5.6)$$

$$p_c = p_c(S_w, a_{wn}), \quad (5.7)$$

where Q_α is the external supply of phase α (e.g. through injection and / or pumping), E_{wn} is the rate of production of specific interfacial area, $\underline{\underline{K}}$ denotes intrinsic permeability, $\underline{\underline{K}}_{wn}$ is the permeability of the wn-interface, and μ_α is dynamic viscosity of the α-phase. As this equation system includes the specific interfacial area a_{wn} as an important new variable, we refer to it as the *2pia model* (**2-p**hase model including **i**nterfacial **a**rea) in the following sections.

5.1.1 Constitutive relationships

Some of the quantities in Eq. (5.1) through (5.7) still depend on other—primary—variables. For those quantities, we need functional dependencies, i.e. constitutive relationships, which relate them to the primary variables. Specifically, these quantities are the specific interfacial area a_{wn} as well as the production/ destruction rate of specific interfacial area E_{wn}. Such relationships could be quite complex. However, because the main goal of this work is to highlight the significance and role of specific interfacial area, we try to use simple relationships when possible.

5.1.1.1 Specific interfacial area

Using the simplified equation system presented in the introduction of Sec. 5.1, capillary pressure is a function of not only saturation as in classical two-phase models, but also a function of interfacial area, see Eq. (5.7). This new functional relationship $p_c(a_{wn}, S_w)$ is generally not single-valued (see Reeves and Celia [1996], Held and Celia [2001], Joekar-Niasar et al. [2008]) meaning that for a certain value of saturation and interfacial area, two different values of capillary pressure are possible.

Alternatively, the capillary pressure–interfacial area–saturation relationship may be re-interpreted as a function relating specific interfacial area a_{wn} to wetting-phase saturation S_w and capillary pressure p_c,

$$a_{wn} = a_{wn}(S_w, p_c). \tag{5.8}$$

This relation is single-valued. It can be obtained by fitting surfaces to a_{wn}-S_w-p_c data coming from either pore-scale network models or from experiments. In this work, computationally generated a_{wn}-S_w-p_c data points are used that were obtained by Joekar-Niasar et al. [2008] using a pore-scale network model. It turned out that the following bi-quadratic relationship generally matches the data excellently (correlation coefficients > 0.95):

$$a_{wn}(S_w, p_c) = \left(a_{00} + a_{10}S_w + a_{01}p_c + a_{20}S_w^2 + a_{11}S_w \cdot p_c + a_{02}p_c^2\right). \tag{5.9}$$

An example of a typical $a_{wn}(S_w, p_c)$ surface is plotted in Fig. 5.1. The projection of all generated a_{wn}-S_w-p_c data points onto the p_c-S_w plane results in the well-known p_c–S_w hysteresis loop, bounded by a p_c^{dr} and a p_c^{imb} function. There exists also interfacial area related to wetting-phase films formed around the solid surface. But the physical behavior of interfaces in these films is completely different from the behavior of mobile interfaces and not governed by capillary forces. Therefore, film interfaces are not included in the formulation of the a_{wn}-S_w-p_c relationship or in the balance equation studied here.

5.1.1.2 Production/ destruction rate of specific interfacial area E_{wn}

We expect the production rate term E_{wn} to be very important for modeling the evolution of interfaces. Unfortunately, there is currently no information available con-

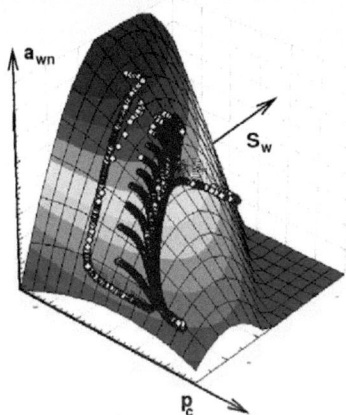

Figure 5.1: Relation between specific interfacial area, saturation, and capillary pressure.

cerning this term. One possibility would be to use a dynamic pore-scale model to generate correlations for this term. In the absence of any other information, we have followed the following physically motivated arguments to construct a formula for E_{wn}.

When a porous medium is fully saturated by a given phase, there are no fluid-fluid interfaces ($a_{wn} = 0$). As the porous medium is invaded by another phase (due to either imbibition or drainage), interfaces are created. Thus, the term E_{wn} will have a positive sign in this range. Obviously, the faster the change in saturation (i.e. for larger $\partial S_w/\partial t$), the larger the rate of change of specific interfacial area (i.e. a larger E_{wn}). As the invasion process continues and the saturation of the invading phase increases, at some point, we reach the ridge saturation S_w^*, i.e. a_{wn} will reach a maximum value. This means that at this point, generation of interfacial area stops, and their destruction starts, i.e. the rate of change of specific interfacial area will be zero ($E_{wn} = 0$). Beyond this point, E_{wn} will have a negative sign, as there would be less and less interfacial area. Obviously, if for any reason the change in phase saturation is halted, there would be no change in interfacial area either.

The foregoing considerations lead us to the following ansatz for the E_{wn} term:

$$E_{wn}(S_w, p_c) = -e_{wn}(S_w, p_c) \cdot \frac{\partial S_w}{\partial t}, \tag{5.10}$$

where e_{wn} is a parameter characterizing the strength of change of specific interfacial area due to a change in saturation.

To complete the parameterization of E_{wn} and e_{wn}, Eq.s (5.10) and (5.3) are combined and it is assumed that the role of interfacial area flux is negligible. Note that this assumption is only made in order to construct the production rate term; in the model, however, the interfacial area flux term is kept. Nevertheless, interfacial area flux is shown to be indeed small in estimations from experimental data by Crandall et al. [2008] who carried out two-phase flow experiments in a stereolithography flow cell with known geometry. Thus, substitution of Eq. (5.10) into the interfacial balance Eq. (5.3), neglecting interfacial flux, and rearranging yields:

$$e_{wn} = \frac{\partial a_{wn}}{\partial p_c} \cdot \left(\frac{\mathrm{d}p_c}{\mathrm{d}S_w}\right)_{\text{line}} + \frac{\partial a_{wn}}{\partial S_w}, \qquad (5.11)$$

where $\frac{\partial a_{wn}}{\partial p_c}$ as well as $\frac{\partial a_{wn}}{\partial S_w}$ can be calculated from Eq. (5.9). The path $\left(\frac{\mathrm{d}p_c}{\mathrm{d}S_w}\right)_{\text{line}}$ is in general unknown. However, along the main (or primary) drainage and imbibition curves, p_c is a known function of S_w and thus, $\left(\frac{\mathrm{d}p_c}{\mathrm{d}S_w}\right)_{\text{line}}$ can be calculated. For all other paths, $\left(\frac{\mathrm{d}p_c}{\mathrm{d}S_w}\right)_{\text{line}}$ is unknown, and therefore e_{wn} has to be obtained by interpolation between the two known values of e_{wn}. This interpolation can be done either with respect to S_w, with respect to p_c, or with respect to both. The interpolation with repect to p_c seems to behave better numerically than the interpolation with respect to S_w, see Faigle [2009],

$$e_{wn}(S_w, p_c) = \frac{e_{wn}^{dr}(S_w, p_c) \cdot \left(p_c - p_c^{imb}(S_w)\right) + e_{wn}^{imb}(S_w, p_c)\left(p_c^{dr}(S_w) - p_c\right)}{p_c^{dr}(S_w) - p_c^{imb}(S_w)}, \qquad (5.12)$$

where the superscripts ^{dr} and ^{imb} indicate main (or primary) drainage and imbibition curve, repectively. Despite this advantage, the restriction of this interpolation is that capillary pressures has to lie within the main hysteresis loop.

This interpolation implies that we expect the potential for change in interfacial area (i.e. e_{wn}) to be the largest at the start of an imbibition or drainage process where the porous medium is fully saturated by one of the phases (and thus, a_{wn} is zero). Note that at the start of main imbibition and drainage, i.e. for both positive and negative $\partial S_w/\partial t$, respectively, interfacial areas are created (i.e. E_{wn} must be positive). This means that e_{wn} has to have a different sign in imbibition and drainage. The values $p_c^{imb}(S_w)$ and $p_c^{dr}(S_w)$ in Eq. (5.12) are known as they relate to the bounding imbibition and drainage curves shown in Fig. 5.1. In fact, if we assume a Brooks-Corey or van Genuchten formula for these curves, we can prescribe $p_c^{\text{imb}}(S_w)$, $p_c^{\text{dr}}(S_w)$, and $\left(\frac{dp_c}{dS_w}\right)_{\text{line}}$ as a function of S_w.

5.1.2 Numerical model

The system of equations presented in Eq.s (5.1) through (5.7) comprises six unknowns: S_w, S_n, p_w, p_n, p_c, and a_{wn}. When numerically solving this system, one has to choose three primary variables; as many as there are partial differential equations. A suitable set is found to be S_w, p_w, and p_c, as the three remaining variables can then be calculated as a unique function of these parameters,

$$S_n = 1 - S_w \qquad (5.13)$$

$$p_n = p_w + p_c \qquad (5.14)$$

$$a_{wn} = a_{wn}(S_w, p_c). \qquad (5.15)$$

For spatial discretization of the three partial differential equations given in Eq. (5.1) through (5.3), we use a fully-coupled vertex-centered finite element method which not only conserves mass locally, but is also applicable to unstructured grids. For time discretization, a fully implicit Eulerian approach is used as this is the easiest and most stable discretization technique for our purpose. The nonlinear system is linearized using a damped inexact Newton-Raphson solver, and the linear system is subsequently solved using a Bi-Conjugate Gradient Stabilized method (known as BiCGStab method), see Bastian et al. [1997].

In the balance equations for both wetting and non-wetting phase, full upwinding is applied to the flux term. For the presented equation system, this means that the relative permeabilities are evaluated with saturations taken from the upstream nodes. This takes accounts of the physical fact that, for the advective part of the equations, information is transfered from upstream to downstream. It also ensures a monotonic behavior of the wetting and non-wetting phase saturations. For the interfacial balance equation, however, no upwinding is applied. Instead, the saturations from the upstream and downstream nodes are arithmetically averaged, thus accounting for the fact that the flux of interfacial area is small—the diffusive term in the interfacial balance equation is small—and that due to the ridge in the $a_{wn}(S_w, p_c)$ surface,

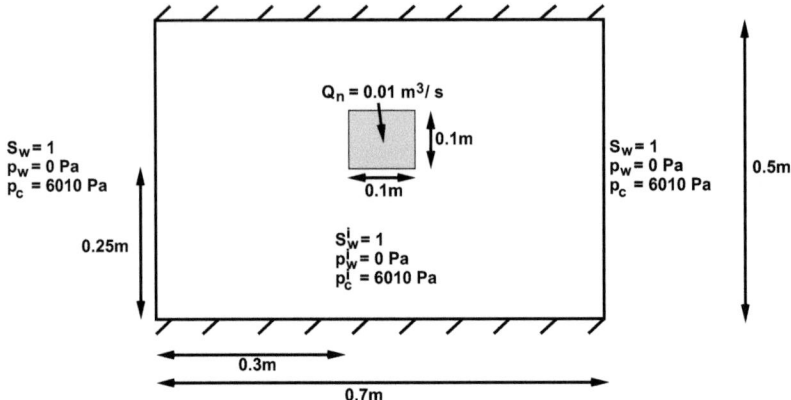

Figure 5.2: Boundary and initial conditions for the drainage problem (example 1).

interfacial area can be non-monotonic (see also Fig. 5.1).

5.1.3 Example: primary drainage

In this example, we simulate the infiltration of a non-wetting phase into a porous medium initially saturated by the wetting phase.

The setup of the example including boundary and initial conditions is shown in Fig. 5.2. The upper and lower boundaries of the rectangular domain are closed to flow of wetting and non-wetting phase as well as to flow of the wn-interface. But the side boundaries are open. The initial values are equal to the boundary values of the left and right boundary. Specifically, the wetting-phase saturation is one, the wetting-phase pressure is set to zero, and the capillary pressure is equal to the entry pressure of 6010 Pa. The boundary and initial values for the 2p model are identical to those given in the 2pia model with the only difference being that no values have to be given for capillary pressure and flux of the wn-interface. We assume that flow occurs in a horizontal plane so that gravity does not play a role.

Parameter values for both 2pia and 2p model are shown in Tab. 5.1. Fluid properties are based on values for water and tetrachloroethene (PCE). The interfacial permeability K_{wn} is assigned a very small value in order to reduce the influence of the diffusive term in Eq. (5.3), as in the absence of experimental data, we do not

	ρ_α $\left[\frac{kg}{m^3}\right]$	μ_α [Pa·s]	K_{wn} $\left[\frac{m^3}{s}\right]$	ϕ [-]	K [m²]	p_d [Pa] (only 2p)	λ [-] (only 2p)
w	998	10^{-3}	10^{-17}	0.3	$3 \cdot 10^{-11}$	6010	5
n	1621	$9 \cdot 10^{-4}$					

Table 5.1: Parameters for the drainage problem of example 1.

know the magnitude of this parameter. The λ value is only needed for the Brooks-Corey parameterization of the capillary-pressure–saturation relationship in the 2p model. This λ value and the entry pressure p_d are obtained by fitting them to the primary drainage capillary pressure–saturation data points of the same pore-scale network model that is used for fitting the constants a_{ij} of the $a_{wn}(S_w, p_c)$ relationship in Eq. (5.9). The constants a_{ij} in turn are obtained by a bi-quadratic fit as $a_{00} = 313.6 \; \frac{1}{m}$, $a_{10} = 5535 \; \frac{1}{m}$, $a_{01} = 0.085 \; \frac{1}{m \cdot Pa}$, $a_{20} = -3937 \; \frac{1}{m}$, $a_{11} = -0.307 \; \frac{1}{m \cdot Pa}$, $a_{02} = -5.2 \cdot 10^{-6} \; \frac{1}{m \cdot Pa^2}$. The contour lines of saturation and capillary pressure obtained by 2p and 2pia model are shown in Fig. 5.3 along with contour lines of specific interfacial area (provided by the 2pia model). These results are for a simulated time $t = 192$ s.

One can see that in both 2p and 2pia model, capillary pressure increases with decreasing wetting-phase saturation. Therefore, the maximum capillary pressure and the minimum wetting-phase saturation can be found in the middle of the domain, inside the source zone of PCE. Contrarily, the maximum of specific interfacial area is located on a ring around that zone. This is due to the fact that the wetting-phase saturation near the source has decreased beyond the ridge value of S_w^* of the a_{wn}–S_w–p_c surface and specific interfacial area has gone down in that zone.

A comparison of the results of the 2pia model and the 2p model shows that for a drainage process only, their is no significant difference. This is due to the fact that hysteresis of the capillary pressure–saturation relationship does not come into play.

5.2 Capillary redistribution

The capillary redistribution of two fluids in a horizontal column filled by a porous medium is an extremely interesting problem as it gives answers to important ques-

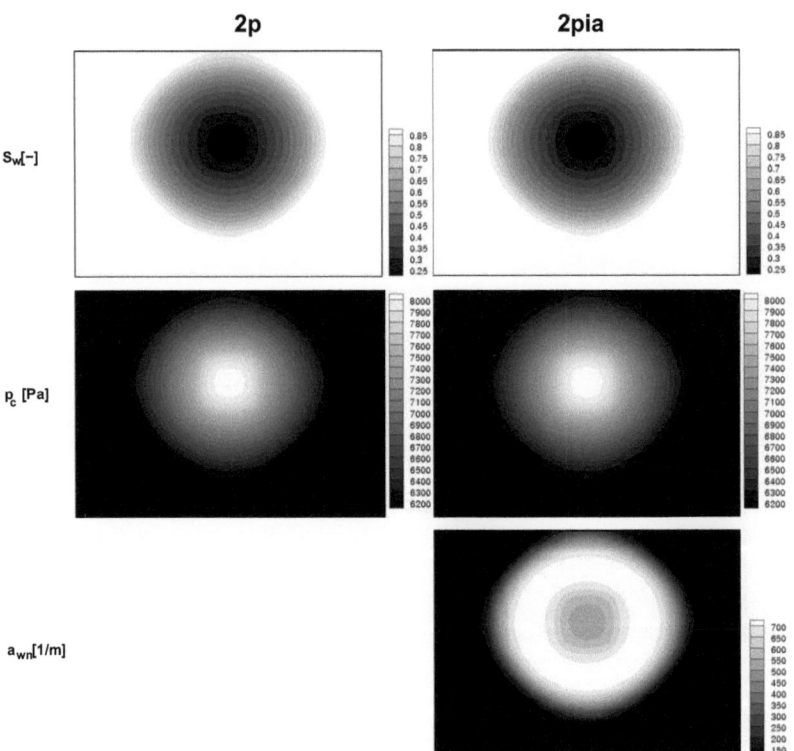

Figure 5.3: Distribution of S_w and p_c in example 1 using the 2p model (left hand side) and of S_w, p_c, and a_{wn} using the 2pia model (right hand side) after 192s.

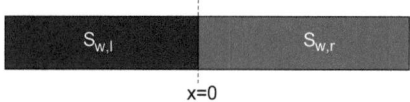

Figure 5.4: Setup of the redistribution problem.

tions on how moisture redistributes in the unsaturated zone of the subsurface. A profound knowledge of the involved processes is essential for water management and irrigation strategies, especially in arid zones.

A simple capillary redistribution problem risen by Philip [1991] is the reference problem for that purpose. He studied the redistribution of air and water in an infinitely long horizontal porous medium with an initial saturation discontinuity. This will be the guiding example through this section. Philip's problem will be addressed in three different ways in the following: mathematically (Sec. 5.2.1), experimentally (Sec. 5.2.2), and numerically (Sec. 5.2.3).

5.2.1 Mathematical approach

Acknowledging the need for macro-scale models accounting for interfaces, it is of crucial interest to find good parameter estimates and also to identify verification possibilities for these models, like the one by Niessner and Hassanizadeh [2008] that was introduced in Sec. 5.1. This can be done by comparison either with experimental data, or with analytical and semi-analytical solutions. In the absence of any experimental data on dynamic evolution of interfaces, we focus on the derivation of a semi-analytical solution for the interfacial area model (interfacial-area-based approach), see Pop et al. [2009] for further details.

For this purpose, we consider the redistribution of two fluid phases in a one-dimensional horizontal porous medium that is homogeneous and infinitely long. This medium is divided into two parts, with the initial wetting-phase saturation being large in one part and small in the other part, see Fig. 5.4. As redistribution takes place, drainage will occur in the high-saturation part of the domain and imbibition will occur on the other side. The two subdomains are separated by an interface where the saturation is discontinuous. Due to the requirement of capillary pressure continuity, this discontinuity in saturation will persist for any positive time. Philip [1991] obtained a semi-analytical solution for a similar problem. In the framework described above, he considered the case of unsaturated flow using the classical model with two clearly defined capillary pressure–saturation curves (one for the drainage subdomain and one for the imbibition subdomain). Assuming the continuity of the flux, as well as of the capillary pressure, Philip obtained a similarity solution for the resulting model.

An analogous construction was proposed in van Duijn and de Neef [1998], where the redistribution problem is studied in the context of an extended capillary pressure condition. The procedure given there is based on the analysis carried out in Atkinson and Peletier [1971, 1974], van Duijn and Peletier [1976/77], where they proved the existence and uniqueness of similarity solutions for the nonlinear (possibly degenerate) diffusion equation, as well as monotonicity properties for the associated fluxes. In the same context, we also mention the work of Fucik et al. [2008]. The semi-analytical solutions given there solve the classical two-phase model including advection terms.

In the study of the redistribution problem, the common practice is to use two different capillary pressure–saturation curves: one for the drainage region and another one for the imbibition region. These curves are assumed known a priori. In our interfacial-area-based approach, we do not make this assumption. Instead we construct such curves from a relationship between interfacial area, saturation and capillary pressure. We apply the procedures described in van Duijn and de Neef [1998] and Philip [1991] to construct similarity solutions for the redistribution problem that includes the interfacial area. This provides a verification tool for numerical models based on the new theory.

The following studies are published in Pop et al. [2009] and are based on the simplified model of Niessner and Hassanizadeh [2008] which is founded on the interfacial-area-based approach and includes phase-interfacial area. In this section, we reduce the mathematical structure of the model given by Eq.s (5.1) through (5.7) to the redistribution problem introduced by Philip [1991], and then apply the procedure in van Duijn and de Neef [1998] for constructing a solution to the resulting problem.

Recalling that the flow is one-dimensional and that the phase-saturations add to unity, the total velocity $v = v_w + v_n$ is constant in space. Here we assume the total velocity being constant in time as well. To model redistribution only, we set $v = 0$. Then, the balance equations (5.1) - (5.2) are reduced to the scalar two-phase flow model

$$\phi\frac{\partial S_w}{\partial t} + K\frac{\partial}{\partial x}\left\{\frac{k_{rw}}{\mu_w}\frac{k_{rn}}{\mu_n}\left(\frac{k_{rw}}{\mu_w} + \frac{k_{rn}}{\mu_n}\right)^{-1}\frac{\partial p_c}{\partial x}\right\} = 0. \quad (5.16)$$

Remark 1 *The same approach can be considered for the case of unsaturated flow, when the non-wetting phase is gaseous (e.g. air) and assumed at a constant pressure, say $p_{air} = 0$. Then we have $p_c = -p_w$ and $v_n = 0$. In this case, the balance equations reduce to*

$$\phi\frac{\partial S_w}{\partial t} + K\frac{\partial}{\partial x}\left\{\frac{k_{rw}}{\mu_w}\frac{\partial p_c}{\partial x}\right\} = 0. \quad (5.17)$$

5.2.1.1 The dimensionless form

In order to construct the similarity solution, we put Eq.s (5.16) and (5.3) in a dimensionless form. Let L, T, P, and A be characteristic values for the length, time, capillary pressure, and interfacial area, respectively, and transform the variables and unknowns by

$$x := \frac{x}{L}, \ t := \frac{t}{T}, \ p := \frac{p_c}{P}, \ a := \frac{a_{wn}}{A}, \ \text{and} \ e := \frac{e_{wn}}{A}. \quad (5.18)$$

We choose T and L such that

$$\frac{T}{L^2} = \frac{\phi\mu_w}{KP}. \quad (5.19)$$

Furthermore, we introduce the viscosity ratio $M = \frac{\mu_n}{\mu_w}$, as well as the interfacial number

$$C_{ia} = \frac{\phi\mu_w A K_{wn}}{KP}. \quad (5.20)$$

Disregarding the subscript in the water saturation S_w, we end up with the dimensionless model

$$\frac{\partial S}{\partial t} + \frac{\partial}{\partial x}\left(\frac{k_{rw}(S)k_{rn}(S)}{Mk_{rw}(S) + k_{rn}(S)}\frac{\partial p}{\partial x}\right) = 0, \quad (5.21)$$

$$\frac{\partial a}{\partial t} - C_{ia}\frac{\partial}{\partial x}\left(a\frac{\partial a}{\partial x}\right) = -e\frac{\partial S}{\partial t}, \quad (5.22)$$

for all $x \in \mathbb{R}$ and $t > 0$. Initial conditions have to be specified for completing the model:

$$S(x,0) = S_0(x) \quad \text{and} \quad p(x,0) = p_0(x) \quad \text{for all} \quad x \in \mathbb{R}. \quad (5.23)$$

Note that the initial conditions are given for the capillary pressure, and not for the interfacial area. Knowing S_0 and p_0, we can use Eq. (5.7) to obtain the initial specific interfacial area a_0.

In view of Remark 1, the dimensionless model for unsaturated flow replaces Eq. (5.21) by

$$\frac{\partial S}{\partial t} + \frac{\partial}{\partial x}\left(k_{rw}(S)\frac{\partial p}{\partial x}\right) = 0. \tag{5.24}$$

Defining the nonlinear function $D(\cdot)$ as

$$D(S) = \begin{cases} \dfrac{k_{rw}(S)k_{rn}(S)}{Mk_{rw}(S) + k_{rn}(S)}, & \text{for the two-phase model,} \\ k_{rw}(S), & \text{for the unsaturated flow model,} \end{cases} \tag{5.25}$$

we can combine both balance equations (5.21) and (5.24) into the single form

$$\frac{\partial S}{\partial t} + \frac{\partial}{\partial x}\left(D(S)\frac{\partial p}{\partial x}\right) = 0. \tag{5.26}$$

In this way we have brought both unsaturated and two phase models to a similar form, the only difference being in the nonlinearity $D = D(S)$.

Based on experimental investigations in flow cells (see Crandall et al. [2008]), one can conclude that the interfacial number is small. It typically ranges between 10^{-6} and 10^{-4} which indicates that the diffusive term in the interfacial area balance equation is much smaller than the two other terms allowing us to disregard this diffusion term. In other words, the parabolic equation (5.22) is replaced by the ordinary differential equation

$$\frac{\partial a}{\partial t} = -e\frac{\partial S}{\partial t}. \tag{5.27}$$

The system is closed by Eq. (5.7) in dimensionless form, i.e. $a = a(S,p)$.

5.2.1.2 The semi-analytical solution

In this section, we discuss the construction of a semi-analytical solution for the model given by Eq.s (5.26) and (5.27). We start by constructing appropriate capillary pressure–saturation curves, and then seek for self-similar solutions in a simplified context.

The capillary pressure–saturation curves Inserting the dimensionless form of Eq. (5.7) into Eq. (5.27) gives

$$\partial_S a \frac{\partial S}{\partial t} + \partial_p a \frac{\partial p}{\partial t} = -e \frac{\partial S}{\partial t}, \qquad (5.28)$$

where $\partial_\alpha a$ denotes partial differentiation with respect to the argument α ($\alpha = S, p$).

As indicated earlier, we assume the existence of an explicit form for the functions $a(S, p)$ and $e(S, p)$. Moreover, using physical reasoning, we restrict our analysis to the range for S and p where the following hold:

$$\partial_p a < 0, \qquad \text{as well as} \qquad e + \partial_S a < 0. \qquad (5.29)$$

Then for any arbitrarily fixed $x \in \mathbb{R}$ we get:

$$\frac{\partial p}{\partial t} = -\frac{e + \partial_S a}{\partial_p a} \frac{\partial S}{\partial t}.$$

Since x is fixed, we can interpret p as a function of S and obtain

$$\frac{dp}{dS} = -\frac{e + \partial_S a}{\partial_p a}. \qquad (5.30)$$

The initial condition associated to Eq. (5.30) is provided by the initial conditions given by Eq. (5.23). Specifically, for the fixed x and at $t = 0$ we have $p = p_0(x)$ and $S = S_0(x)$, implying

$$p(S_0) = p_0. \qquad (5.31)$$

In this way, we end up with a family of initial value problems depending on the parameter $x \in \mathbb{R}$:

$$\frac{dp}{dS} = -\frac{e + \partial_S a}{\partial_p a}, \qquad (5.32)$$

$$p(S_0(x)) = p_0(x). \qquad (5.33)$$

Solving these problems for any $x \in \mathbb{R}$ would provide the capillary pressure–saturation curve at that point, $p(x, t) = p(S(x, t); x)$. Notice that both t and x are acting only as parameters.

In view of Eq. (5.29), the functions on the right hand side in Eq. (5.32) are smooth in both arguments S and p. Standard theory for ordinary differential equations ensures the existence and uniqueness of a solution for the problem defined by Eq.s (5.32) through (5.33). This property has two immediate consequences. First, any pair (S, p) defines a unique capillary pressure–saturation curve satisfying $p(S) = p$. In particular, if the pair (S, p) lies on a bounding capillary pressure–saturation curve (drainage or imbibition), then drainage or imbibition will follow along that curve. Furthermore, the uniqueness for Eq.s (5.32) through (5.33) also implies that two different p–S curves can never intersect. In particular, a secondary p - S curve cannot cross any of the two bounding curves, and therefore it will stay inside the domain defined by these. In Sec. 5.2.1.3 below, we take this into account when constructing the production function e.

Next, the existence and uniqueness of a solution also shows a limitation of the ansatz in Eq. (5.10) for the interfacial area production term. Specifically, there is a unique capillary pressure–saturation curve through a given point, regardless of the process taking place: drainage or imbibition. In other words, the assumed form for E and e does not make any distinction between drainage and imbibition. A possible remedy is to let e depend also on the sign of $\partial_t S$, allowing then for two different p - S curves corresponding to either drainage or imbibition processes. However, here we seek for a solution to the redistribution problem, where drainage is encountered in one subdomain, whereas imbibition appears in the other one. Therefore, the shortcoming of Eq. (5.10) has no effect in the present context.

Remark 2 *For defining the problem of Eq.s (5.32) through (5.33), we have used the interfacial area equation (5.27). Consequently, along any capillary pressure–saturation curve solving Eq.s (5.32) through (5.33), Eq. (5.27) will be satisfied automatically. We will use this remark in the next section, where the interfacial area equation (5.27) is disregarded when constructing a solution to our model.*

The similarity solution In this section we construct a self-similar solution of the interfacial area model by considering special initial data for saturation and capillary pressure. Specifically, we assume that initially, both S and p are constant to the left and to the right of $x = 0$:

$$S_0(x) = \begin{cases} S_\ell, & \text{if } x < 0, \\ S_r, & \text{if } x > 0, \end{cases} \quad \text{and} \quad p_0(x) = \begin{cases} p_\ell, & \text{if } x < 0, \\ p_r, & \text{if } x > 0. \end{cases} \quad (5.34)$$

Notice that in each subdomain, both the initial saturation and initial pressure are constant. However, a pressure gradient is encountered at the interface $x = 0$, which causes flow from one subdomain to the other one. As a consequence, drainage takes place in one of the subdomains and imbibition in the other one. Without loss of generality, we assume that the initial conditions are chosen such that drainage occurs for $x < 0$ and imbibition for $x > 0$.

As follows from Eq.s (5.32) through (5.33), only two capillary pressure–saturation curves have to be computed: $p^- = p^-(S)$, obtained for the initial data $p^-(S_\ell) = p_\ell$ for all $x < 0$, and $p^+ = p^+(S)$ with $p^+(S_r) = p_r$ for all $x > 0$. In view of Remark 2, along any of these curves, the interfacial area equation (5.27) is satisfied automatically. Thus having determined the two curves $p^\pm(S)$, we can disregard Eq. (5.27) and reduce the original model to

$$\begin{cases} \dfrac{\partial S}{\partial t} + \dfrac{\partial}{\partial x}\left(D(S)\dfrac{\partial p^-(S)}{\partial x}\right) = 0, \\ S(x,0) = S_\ell, \end{cases} \qquad (5.35)$$

for $x < 0$, respectively

$$\begin{cases} \dfrac{\partial S}{\partial t} + \dfrac{\partial}{\partial x}\left(D(S)\dfrac{\partial p^+(S)}{\partial x}\right) = 0, \\ S(x,0) = S_r, \end{cases} \qquad (5.36)$$

for $x > 0$.

The interface conditions The two submodels are coupled at the interface $x = 0$, where we impose continuity of water flux and pressure. Introducing the right and the left limits of the saturation at $x = 0$:

$$S^-(t) = S(0-,t), \quad \text{and} \quad S^+(t) = S(0+,t), \qquad (5.37)$$

the coupling conditions become

$$D(S^-(t))\dfrac{\partial p^-(S^-(t))}{\partial x} = D(S^+(t))\dfrac{\partial p^+(S^+(t))}{\partial x}, \qquad (5.38)$$

and

$$p^-(S^-(t)) = p^+(S^+(t)). \qquad (5.39)$$

Remark 3 *In this framework, we have reduced the interfacial area model of Eq.s (5.26) and (5.27) to the redistribution problem posed by (Philip [1991]). The major difference here is due to the interfacial area. Whereas in the Philip problem two pressure–saturation curves are specified a priori (one for drainage, respectively one for imbibition), we use the interfacial area and the initial data to determine these curves in the present approach.*

We proceed by seeking for self-similar solutions of Eq.s (5.35) and (5.36). As in Atkinson and Peletier [1971, 1974], van Duijn and de Neef [1998], van Duijn and Peletier [1976/77] as well as Philip [1991] we set

$$S = S(\eta) \quad \text{and} \quad p = p(\eta) \quad \text{with} \quad \eta = \frac{x}{\sqrt{t}}.$$

and obtain the two subproblems (SP_ℓ) and (SP_r)

$$(SP_\ell) \begin{cases} -\frac{\eta}{2}S' + \left(D(S)\, p^-(S)'\right)' = 0, & \text{for } \eta \in (-\infty, 0), \\ S(-\infty) = S_\ell, \end{cases} \quad (5.40)$$

and

$$(SP_r) \begin{cases} -\frac{\eta}{2}S' + \left(D(S)\, p^+(S)'\right)' = 0, & \text{for } \eta \in (0, +\infty), \\ S(+\infty) = S_r. \end{cases} \quad (5.41)$$

The two subproblems are coupled at the interface $\eta = 0$ by the conditions given in Eq.s (5.38) and (5.39). To express these conditions in terms of the similarity variable η, we use the unknown limits

$$S^- = \lim_{\eta \nearrow 0} S(\eta), \quad \text{and} \quad S^+ = \lim_{\eta \searrow 0} S(\eta), \quad (5.42)$$

and define the left and right fluxes at $\eta = 0$:

$$\begin{aligned} F^-(S^-) &= D(S^-)\frac{dp^-}{dS}\Big|_{S=S^-} \lim_{\eta \nearrow 0} S'(\eta), \\ F^+(S^+) &= D(S^+)\frac{dp^+}{dS}\Big|_{S=S^+} \lim_{\eta \searrow 0} S'(\eta). \end{aligned} \quad (5.43)$$

In this way the conditions at the interface become

$$F^-(S^-) = F^+(S^+), \quad \text{respectively} \quad p^-(S^-) = p^+(S^+). \quad (5.44)$$

The left and right limits $S^{\pm}$ The main step in solving the subproblems $(SP_{\ell,r})$ is to determine the left and right limits $S_{\pm}$ of the saturation at $\eta = 0$. To do so, we follow the construction proposed in van Duijn and de Neef [1998] for the redistribution problem in the case of a porous column involving two different permeabilities. The procedure is based on results for similarity solutions in semi-infinite as well as infinite domains obtained in Atkinson and Peletier [1971, 1974] as well as by van Duijn and Peletier [1976/77]. We start mentioning the following

Theorem 1 (Existence and uniqueness):
Given $S^- < S_\ell$, Problem (SP_ℓ) with $S(0-) = S^-$ has a unique solution, which is decreasing. Similarly, any $S^+ > S_r$ uniquely determines a (decreasing) solution to Problem (SP_r) satisfying $S(0+) = S^+$.

As proven in the papers mentioned above, an $\eta_l \in [-\infty, 0)$ exists defining a maximal interval $(\eta_\ell, 0)$ on which S is strictly decreasing. Furthermore, if $\eta_\ell > -\infty$, then $S(\eta) = S_\ell$ for all $\eta \leq \eta_\ell$. Analogously, S is strictly decreasing on $(0, \eta_r)$, whereas $S(\eta) = S_r$ for all $\eta \geq \eta_r$ whenever $\eta_r < \infty$. Necessary and sufficient conditions for the finiteness of η_ℓ and η_r are given in Atkinson and Peletier [1971, 1974], van Duijn and Peletier [1976/77].

Based on Theorem 1, we discuss the initial conditions given in Eq. (5.34). As assumed in the beginning of Sec. 5.2.1.2, drainage occurs for $x(\eta) < 0$ and imbibition for $x(\eta) > 0$. Since the solution is decreasing in both subdomains and since both pressure–saturation curves $p^{\pm}$ are decreasing, the pressure continuity at $\eta = 0$ gives:

$$p^-(S_\ell) \leq p^-(S^-) = p^+(S^+) \leq p^+(S_r).$$

The case $p^-(S_\ell) = p^+(S_r)$ is trivial since the solution is constant in each subdomain. Therefore we restrict ourselves to the cases when the inequalities in the above are strict, yielding a necessary condition for the solvability of the coupled $(SP_{\ell,r})$ problems:

$$p^-(S_\ell) < p^+(S_r). \tag{5.45}$$

Furthermore, $p^{\pm}$ are strictly monotone, thus invertible. We introduce the strictly increasing functions $h_-^+, h_+^- : [0,1] \to [0,1]$

$$h_-^+(s) := (p^+)^{-1}(p^-(s)), \quad \text{respectively} \quad h_+^-(s) := (p^-)^{-1}(p^+(s)), \tag{5.46}$$

for any $s \in [0,1]$. Then, the pressure continuity at $\eta = 0$, as well as Eq. (5.45), become

$$S^+ = h_-^+(S_-), \quad \text{or} \quad S^- = h_+^-(S_+), \quad \text{respectively} \tag{5.47}$$

$$S_\ell > h_+^-(S_r), \quad \text{or} \quad S_r < h_-^+(S_\ell).$$

Using this notation, we can reformulate the results by van Duijn and Peletier [1976/77]:

Theorem 2 (Monotonicity of $F^\pm$):
The flux F^- is increasing in S^- for any $h_+^-(S_r) \leq S^- \leq S_\ell$. Furthermore,

$$F^-(S_\ell) = 0, \quad \text{and} \quad F^-(h_+^-(S_r)) > 0.$$

Similarly, F^+ is decreasing in S^+ for any $h_-^+(S_\ell) \geq S^- \geq S_r$, with

$$F^+(S_r) = 0, \quad \text{and} \quad F^+(h_-^+(S_\ell)) > 0.$$

Both fluxes $F^\pm$ depend continuously on the arguments.

Theorem 2 immediately implies the existence and uniqueness of a solution pair (S^-, S^+) satisfying the interface conditions given by Eq. (5.44). To see this, we first use the pressure condition for writing $S^+ = h_-^+(S^-)$ and seek for an $S^- \in [h_+^-(S_r), S_\ell]$ yielding the flux continuity at the interface. With

$$\mathcal{F} : [h_+^-(S_r), S_\ell] \to \mathbb{R}, \quad \mathcal{F}(s) = F^-(s) - F^+(h_-^+(s)), \tag{5.48}$$

we have defined a continuous and increasing function satisfying

$$\begin{aligned} \mathcal{F}(S_\ell) &= F^-(S_\ell) - F^+(h_-^+(S_\ell)) < 0, \quad \text{and} \\ \mathcal{F}(h_-^+(S_r)) &= F^-(h_-^+(S_r)) - F^+(S_r) > 0, \end{aligned} \tag{5.49}$$

implying the existence of a unique $s_0 \in (h_+^-(S_r), S_\ell)$ solving $\mathcal{F}(s_0) = 0$. Taking $S^- = s_0$ and $S^+ = h_-^+(s_0)$ we have obtained the left and right limits of the self-similar saturation at the interface $\eta = 0$.

The above existence result suggests a straightforward numerical approach for calculating the saturation pair (S^-, S^+), the bisection. Such an approach uses, however, the fluxes $F^\pm(\cdot)$ evaluated for different arguments. In a direct approach, given a saturation pair $(\underline{s}, \overline{s})$ satisfying the pressure condition $\overline{s} = h^+_-(\underline{s})$ (see Eq. (5.47)), one can use Eq. (5.43) to compute $F^-(\underline{s})$ and $F^+(\overline{s})$. This requires solving Eq. (5.40$_1$) on the semi-infinite intervals $(-\infty, 0)$ and with $S(-\infty) = S_\ell$, $S(0-) = \underline{s}$, as well as the similar equation (5.41$_1$) on $(0, \infty)$, where $S(0+) = \overline{s}$ and $S(\infty) = S_r$. Notice that in this way, we determine the saturation S on the entire real axis and for any pair of values $(\underline{s}, \overline{s})$, whereas only $F^-(\underline{s})$ and $F^+(\overline{s})$ are needed. The solution S can be computed later and only for the correct values $S^\pm$.

To reduce the computational effort, we consider the approach in van Duijn and de Neef [1998] that is based on the monotonicity of S in η. For any given pair $(\underline{s}, \overline{s} = h^+_-(\underline{s}))$ we can invert $S : (-\eta_\ell, 0) \to (\underline{s}, S_\ell)$ and define $\eta : (\underline{s}, S_\ell) \to (-\eta_\ell, 0)$. This gives $S'(\eta(S)) = 1/\eta'(S)$, yielding $F^-(S) = D(S)\frac{dp^-}{dS}\frac{1}{\eta'(S)}$. Therefore, Eq. (5.40$_1$) can be transformed into

$$\frac{d}{dS}(F^-(S)) = \frac{\eta}{2}, \quad \text{for } \underline{s} < S < S_\ell. \tag{5.50}$$

In the limit $S \searrow \underline{s}$ we get $\frac{dF^-}{dS}(S = \underline{s}) = 0$, whereas $F^{-'}(S_\ell) = 0$ by Theorem 2. Using these conditions and differentiating Eq. (5.50) in S gives the flux problems (FP_ℓ) and (FP_r)

$$(FP_\ell) \begin{cases} F^-(S)\dfrac{d^2}{dS^2}(F^-(S)) = \dfrac{1}{2}D(S)\dfrac{dp^-}{dS}, & \text{for } \underline{s} < S < S_\ell, \\ \dfrac{dF^-(\underline{s})}{dS} = 0, & F^-(S_\ell) = 0. \end{cases} \tag{5.51}$$

In a similar way, on the positive subinterval, we have

$$(FP_r) \begin{cases} F^+(S)\dfrac{d^2}{dS^2}(F^+(S)) = \dfrac{1}{2}D(S)\dfrac{dp^+}{dS}, & \text{for } S_r < S < \overline{s}, \\ \dfrac{dF^+(\overline{s})}{dS} = 0, & F^+(S_r) = 0. \end{cases} \tag{5.52}$$

The problems above can be solved for any pair of values $(\underline{s}, \overline{s})$ satisfying the pressure condition, yielding the left and right fluxes $F^-(\underline{s})$ and $F^+(\overline{s})$. In view of Eq. (5.49), there exists a unique pair yielding the flux continuity at $\eta = 0$. This pair also represents the sought saturation limits $S^\pm$. To determine these two values, we consider the following algorithm, which is based on the conditions at the drainage–

imbibition interface given by Eq. (5.44), as well as Theorem 2.

Determine the curves $p^{\pm}(\cdot)$ by solving the problem defined by Eq.s (5.32) and (5.33) with the initial data given in Eq. (5.34).

Let $S_< := h_+^-(S_r)$ and $S_> := S_\ell$.

While $(S_> - S_< > \varepsilon)$ do (ε being a given tolerance)

$\underline{s} := (S_> + S_<)/2$.

Solve Problem (FP_ℓ), giving $F^-(\underline{s})$.

Solve Problem (FP_r) with $\overline{s} = h_-^+(\underline{s})$, giving $F^+(\overline{s})$.

Compute $\mathcal{F}(\underline{s})$ as given in Eq. (5.48). If $\mathcal{F}(\underline{s}) > 0$ let $S_< := \underline{s}$. If $\mathcal{F}(\underline{s}) < 0$ let $S_> := \underline{s}$.

Go to the beginning of the "while" loop.

Take $S^- = \underline{s}$ and $S^+ = h_-^+(\underline{s})$.

In the above algorithm, we have $S_> \geq S_<$. This is ensured by the Theorems 1 and 2, as well as by the monotonicity of h_-^+ and h_+^-.

Having determined $S^\pm$, we can find a solution to the problems $(SP_{\ell,r})$. To do so, we follow van Duijn and de Neef [1998] and use the fluxes $F^\pm$ solving the problems $(FP_{\ell,r})$. Specifically, Eq. (5.50) immediately gives

$$\eta(S) = \begin{cases} 2\dfrac{d}{dS}\left(F^+(S)\right) > 0, & \text{if } S_r < S < S^+, \\ 2\dfrac{d}{dS}\left(F^-(S)\right) < 0, & \text{if } S^- < S < S_\ell. \end{cases} \quad (5.53)$$

Clearly, the solution $S : \mathbb{R}\setminus\{0\} \to (S_r, S^+) \cup (S^-, S_\ell)$ is determined by inverting η given above. At $\eta = 0$, this solution is completed by the left and right limits $S(0-) = S^-$ and $S(0+) = S^+$.

5.2.1.3 An example

In this section, we present an example for the redistribution problem including interfacial area effects. For the numerical calculations we have considered two fluids having equal viscosities, yielding $M = 1$. The relative permeabilities are assumed quadratic as this result is suggested by the thermodynamically consistent approach. In the dimensionless framework, they become $k_{rw}(S) = S^2$, and $k_{rn}(S) = (1-S)^2$. Consequently, the nonlinear diffusion coefficient in Eq. (5.25) becomes $D(S) = S^2(1-S)^2/(S^2 + (1-S)^2)$. For the bounding capillary pressure–saturation curves, we have chosen power functions with negative and sub-unitary arguments: $p^{dr}(S) = S^{-\frac{1}{\lambda_D}}$ (according to the Brooks–Corey parametrization) with $\lambda_D = 2$ for the bounding drainage curve, and $p^{imb}(S) = S^{-\frac{1}{\lambda_I}}$ with $\lambda_I = 4$ for the bounding imbibition curve (see Fig. 5.5).

For the dimensionless form of the interfacial area equation (5.7), we consider a bi-quadratic dependency on S and p,

$$a = a(S,p) = a_{00} + a_{10}S + a_{01}p + a_{20}S^2 + a_{02}p^2 + a_{11}Sp. \tag{5.54}$$

This form has been suggested based on numerical studies using a pore network model (see Joekar-Niasar et al. [2008]), where appropriately chosen coefficients have led to a good agreement with the numerical data. However, the investigations in Joekar-Niasar et al. [2008] are focussed on an intermediate range of values for the saturation, excluding the fully saturated and the completely unsaturated regimes. In the dimensionless setting, the parameters determined there become

$$a_{00} = -0.036, a_{10} = 0.55, a_{01} = 0.051,$$

$$a_{20} = -0.39, a_{11} = -0.184, \text{ and } a_{02} = -0.019.$$

As revealed by experiments and physical reasoning (see Culligan et al. [2004, 2006]), the interfacial area should go to zero for $S = 0$ and $S = 1$. This is, however, not satisfied with these parameter values because the fitting of the surface a–S–p to the data was done for an intermediate regime. Therefore, the parameterization is not accurate in these extreme situations. In spite of this, we prefer using the present set of coefficients instead of constructing an artificial surface that meets the endpoints (i.e. $S = 0$ or $S = 1$) accurately. The example below also avoids the fully unsaturated or the completely saturated regimes. In this case, we have $\partial_p a(S,p) < 0$ for any pair (S, p) inside the region of the S–p plane that is determined by the bounding capillary pressure–saturation curves. Furthermore, it is less important for our analysis whether a becomes 0 or not in a certain point as the present approach only involves the derivatives of a and not its value. Therefore changing a_{00} in the above formula will have no effect on the solution.

Unfortunately, neither experimental nor numerical data is available yet for the production function e in Eq. (5.27). In choosing a particular form for e one should account for two important situations. These occur when the saturation–pressure pair is moving along the bounding drainage function $p^{dr}(S)$ and along the bounding imbibition function $p^{imb}(S)$. Since these curves are bounding, the model should rule out the situation that the (S,p) pair leaves the domain in the $S-p$ plane which is comprised between these curves. Proceeding as in Eq. (5.28), we obtain

$$e(S,p) = -\partial_p a(S,p)\frac{dp}{dS}(S) - \partial_S a(S,p) \tag{5.55}$$

along the known curves $p^{dr}(\cdot)$ and $p^{imb}(\cdot)$. To extend the definition of e to the entire region of interest we use linear interpolation along constant saturation. Specifically, given a pressure p we can determine uniquely two saturation values S^{dr} and S^{imb} satisfying $p^{dr}(S^{dr}) = p$, respectively $p^{imb}(S^{imb}) = p$. This defines

$$e^{dr}(p) = -\partial_p a(S^{dr},p)\frac{dp^{dr}}{dS}(S^{dr}) - \partial_S a(S^{dr},p), \quad \text{respectively}$$

$$e^{imb}(p) = -\partial_p a(S^{imb},p)\frac{dp^{imb}}{dS}(S^{imb}) - \partial_S a(S^{imb},p).$$

Interpolating between the two curves we define

$$\bar{e}(S,p) = \frac{e^{imb}(p) - e^{dr}(p)}{S^{imb} - S^{dr}}(S - S^{dr}) + e^{dr}(p). \tag{5.56}$$

At this point, we remark that along any curve, the capillary pressure is a decreasing function of the saturation. Recalling Eq. (5.30) and since $\partial_p a < 0$ the sum $e + \partial_S a$ should remain negative for any point (S,p) between the two bounding curves. Observe that this condition is satisfied automatically along these curves. Therefore, we define

$$e(S,p) = \min\{-\partial_S a, \bar{e}(S,p)\}. \tag{5.57}$$

Initially, we assume the following values for the left and right subdomains

$$S_\ell = 0.8 \text{ and } p_\ell = 1.10, \text{ respectively } S_r = 0.2 \text{ and } p_r = 1.68.$$

We have used this data to determine the corresponding drainage, respectively imbi-

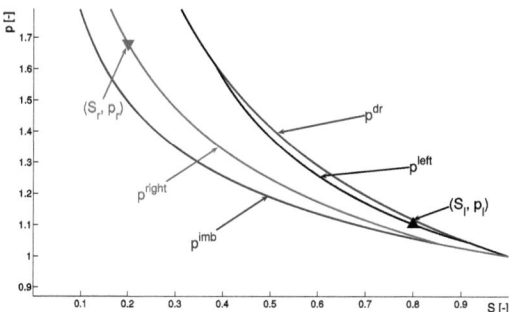

Figure 5.5: The capillary pressures: the outer left and right curves are the bounding imbibition and drainage curves. The interior curves are the ones corresponding to the initial data.

bition curves. These are (numerical) solutions to the problem defined by Eq.s (5.32) through (5.33), where we have used the expressions in Eq.s (5.55) and (5.57) for the interfacial area, respectively the production rate. The outer curves in Fig. 5.5 are the bounding capillary pressure–saturation curves. The interior left curve is the capillary pressure–saturation curve passing through (S_r, p_r) in the imbibition subdomain $\eta > 0$. Similarly, the interior right curve corresponds to the initial pair (S_ℓ, p_ℓ) in the drainage subdomain $\eta < 0$.

The fluxes in Fig. 5.6 are numerical solutions to the problems (FP_r), respectively (FP_ℓ). The right and left saturation limits at $\eta = 0$ are providing continuity for the capillary pressure $(p^-(S^-) = p^+(S^+))$, as well as for the flux across the interface. Applying the scheme described in the previous section, we have obtained $S^+ = 0.365$ and $S^- = 0.498$. Correspondingly, at the interface we obtain $F^- = F^+ = 0.065$, respectively $p^- = p^+ = 1.38$.

Fig. 5.7 displays the saturation in the similarity coordinate $\eta = x/\sqrt{t}$. Drainage occurs in the left subdomain, where the saturation decreases from the initial value $S_\ell = 0.8$ to $S^- = 0.498$. On the right, imbibition is encountered, with S decreasing from $S^+ = 0.365$ to $S_r = 0.2$.

Finally, the pressure is displayed in Fig. 5.8 in the similarity coordinate η. Notice the continuity of the pressure at the interface. Furthermore, since the saturation is decreasing in both subdomains, the capillary pressure is increasing from $p_\ell = 1.10$

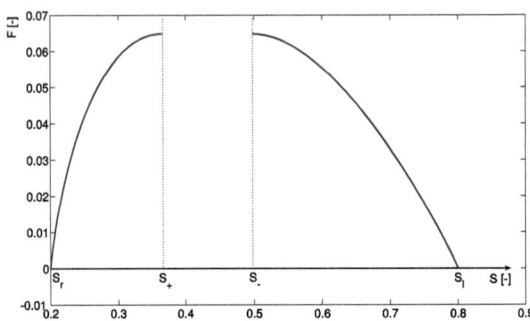

Figure 5.6: The imbibition (left) and drainage (right) fluxes as functions of the saturation. Here $S^+ = 0.365$ and $S^- = 0.498$.

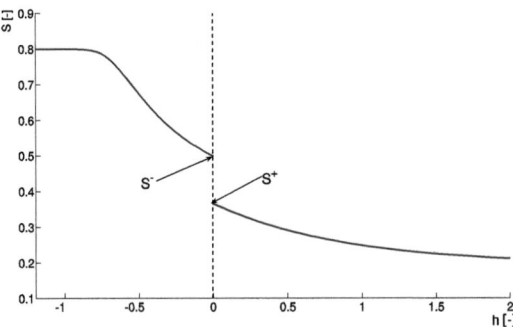

Figure 5.7: The selfsimilar saturation in the imbibition (left) and drainage (right) subdomains. Here $S_\ell = 0.8$, $S^- = 0.498$, $S^+ = 0.365$ and $S_r = 0.2$.

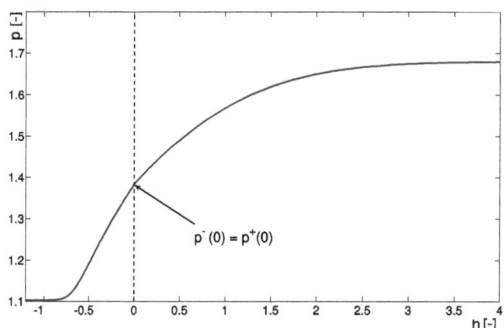

Figure 5.8: The selfsimilar capillary pressures in the imbibition (left) and drainage (right) subdomains. Here $p_\ell = 1.10$, $p^- = p^+ = 1.38$ and $p_r = 1.68$.

to $p_r = 1.68$.

5.2.2 Experimental approach

In order to investigate the "true" solution to the redistribution problem, Braun, Hassanizadeh, Niessner, and Hilfer are currently studying the problem experimentally. The main challenges for experimentalists, when studying setups such as Philip's problem, are: 1) to provide a discontinuous initial condition in a homogeneous porous medium and 2) to accurately measure capillary pressure and saturation as a function of space and time.

A major difference between semi-analytical solution and experiment is that experimentally, it is impossible to provide an infinite domain and thus, investigation of the saturation profile for $t \to \infty$ will have to be considered differently. Mathematically, in an infinite domain, a saturation discontinuity at the drainage-imbibition interface will remain, see Fig. 5.9. Once one of the fronts reaches a domain boundary in the experiment, the flow process will change and will no longer follow a single drainage and a single imbibition curve as indicated in Fig. 5.9. Instead, the saturation discontinuity will be diminished stepwise as indicated in Fig. 5.10—the numbers show different points in time. In any case, the current theories prescribe a constant pressure distribution throughout the column at the end of the experiment. The thermodynamically consistent theory of two-phase flow suggests that it is possible to have

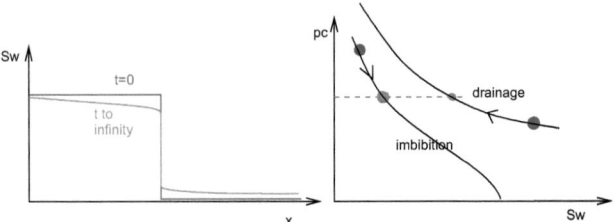

Figure 5.9: Saturation distribution and capillary pressure–saturation curves in an infinite porous medium. Initial values and situation for $t \to \infty$.

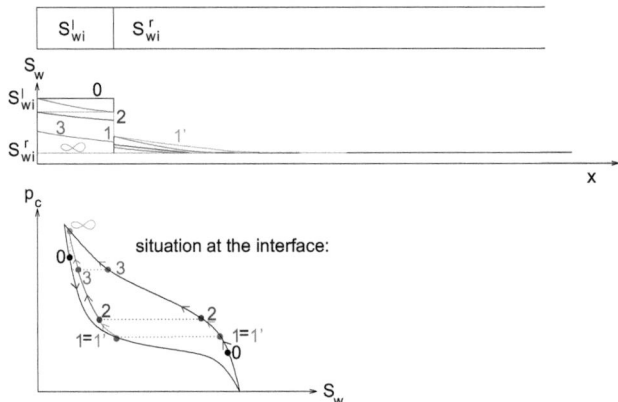

Figure 5.10: Saturation distribution and capillary pressure–saturation curves in a semi-infinite or finite porous medium.

a pressure gradient as well as a saturation gradient under equilibrium conditions. Measurements of pressure should easily determine which theory is correct. So, the goal of the experiment is to study the highly complex processes occurring during the horizontal redistribution of two fluid phases in a porous medium depending on capillary pressure and saturation. The possibilities for the presence of additional terms in Darcy's law will also be investigated.

For the experiment, a Plexiglas column (L = 200 cm, ID = 3 cm) will be packed with sand (already wetted to a desired saturation) to a predetermined level, and sealed with a very thin sheet metal tongue. Then the remainder of the column will be packed with a lower-saturation sand. Of course, it needs to be realized that the position of this tongue determines the initial amount of water in the system and hence, it will eventually control the final capillary pressure in the system. Once the whole column is filled, it will be placed horizontally in the gamma-system shown in Fig. 5.11. Due to the horizontal placement and the small diameter of 3 cm, gravitational effects may be neglected. From the volume of the column and the measured mass of sand and water, packing parameters (density, porosity, initial saturation in the system) will be determined. At predefined distances from the drainage–imbibition interface pressure transmitters will be installed to determine both the wetting phase and the non-wetting phase pressure. In order to deduce the capillary pressure, water-wet and air-wet transmitters will be placed exactly opposite from each other. The transmitters as well as the gamma system will be started to take initial readings. The gamma measurements will also be an indication of the homogeneity of the packing. Then, the steel tongue will be removed and water will start draining/imbibing. The gamma source will be moved along the column to measure the saturation. The gamma-robot will follow a pre-set time–space scheme. The accuracy of the saturation measurement is a strong function of the measuring time. The longer the time span, the more accurate is the reading. In order to capture the dynamics of the systems in the initial phase, trade-offs have to be made. Either few (specific) points can be measured with a high accuracy or more points might be measured with a lesser accuracy. (In order to obtain a high accuracy, a time interval of 5–10 minutes has to be expected).

Measurement results will allow to compare numerical and experimental results to the semi-analytical solution. Additionally, they may help to answer the essential question whether saturation gradients may be a driving force for flow besides pressure gradients and gravity.

5.2.3 Numerical approach

The numerical solution of the redistribution problem is very challenging as the correct saturation jump across the drainage–imbibition interface needs to be guaranteed. This discontinuity cannot be represented using a classical numerical scheme. In a Dipl.-Ing. thesis by Marshall [2009], this problem was addressed by coupling two subdomains in which the dimensionless redistribution equations given by Eq.s (5.21), (5.22), and (5.7) are solved. Specifically, the above set of equations for the drainage subdomain and for the imbibtion subdomain were coupled via an interface condition that allows to guarantee continuity conditions across the drainage–

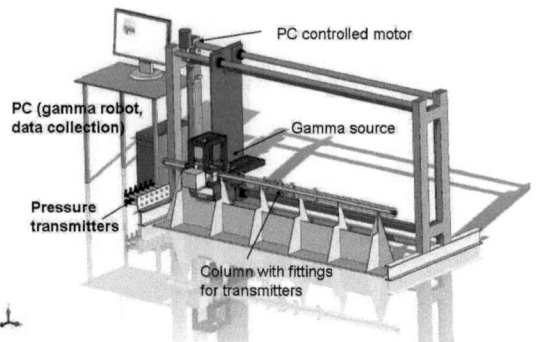

Figure 5.11: Gamma-system with robot for horizontal movement and measuring column. Measuring range of gamma robot: ≈180 cm (figure: VEGAS).

imbibition interface and to correctly represent the saturation jump across that interface. In this respect, Marshall [2009] investigated both the case where the interfacial area number C_{ia} is zero and the case where it is larger than zero. While the first case can be compared to the semi-analytical solution the second case is more relevant in practice. The continuity requirements depend on the character of the interfacial area equation which is determined by the parameter C_{ia}. For the case $C_{ia} = 0$, the following continuity conditions must hold at the drainage–imbibition interface:

- continuity of capillary pressure $p^- = p^+$ and
- continuity of wetting-phase flux, $F^- = F^+$.

For the case $C_{ia} \neq 0$, the interfacial area balance equation is of diffusive type. Therefore, these two continuity conditions need to be supplemented by two additional continuity conditions:

- continuity of specific interfacial area $a^- = a^+$ and
- continuity of interface flux, $F^-_{wn} = F^+_{wn}$.

Keeping this in mind, Marshall [2009] used the following algorithm:

1. Set initial conditions. If $x < 0$: $S = S^l$ and $p = p^l$. Else: $S = S^r$ and $p = p^r$.

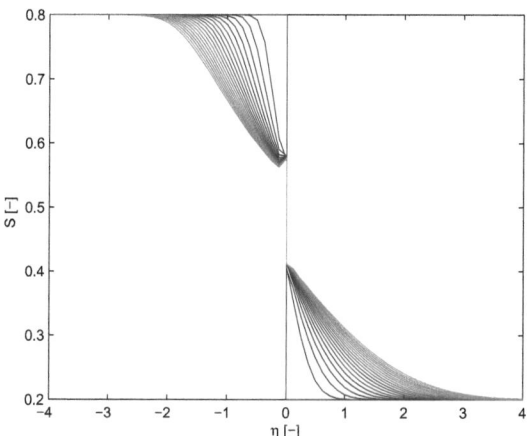

Figure 5.12: Profiles of dimensionless saturation $S(\eta)$ over time.

2. Solve the set of continuity conditions at the interface $\Rightarrow S^+, S^-, p$.

3. Solve the left subproblem using S^- and p as Dirichlet boundary conditions on the right side boundary.

4. Solve the right subproblem using S^+ and p as Dirichlet boundary conditions on the left hand boundary.

5. Based on the results of steps 3 and 4, calculate the water fluxes F^- and F^+ (as well as F_{wn}^- and F_{wn}^+ if $C_{ia} \neq 0$).

6. If $\left|\frac{F^- - F^+}{F^+}\right| < \varepsilon$ (and $\left|\frac{F_{wn}^- - F_{wn}^+}{F_{wn}^+}\right| < \varepsilon$ in case $C_{ia} \neq 0$) with ε as a set limit: converged, go to next time step.
Else: iterate, go to step 2.

Using this algorithm, Marshall [2009] obtained the profiles of dimensionless wetting-phase saturation, capillary pressure, and specific interfacial for a simulation with $C_{ia} = 0$ as shown in Fig.s 5.12 through 5.14. It can be clearly seen that starting from an initial step function, saturation, capillary pressure, and specific interfacial area profiles develop over time in both subdomains. Most importantly, the figures show that the algorithm allows to maintain a discontinuity in saturation and specific interfacial area while having a continuous capillary pressure. The jumps in

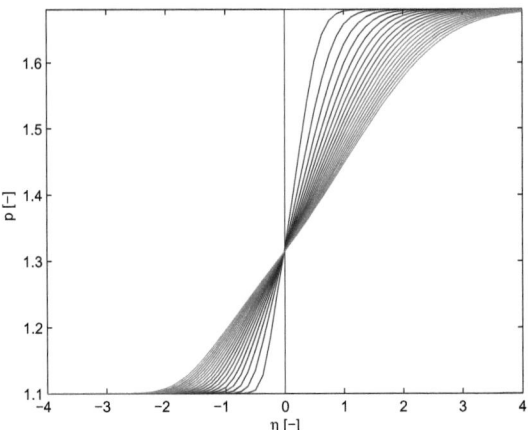

Figure 5.13: Profiles of dimensionless capillary pressure $p(\eta)$ over time.

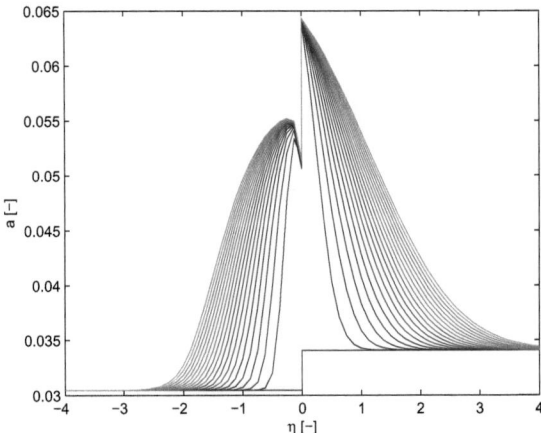

Figure 5.14: Profiles of dimensionless specific interfacial area $a(\eta)$ over time.

saturation and specific interfacial area are such that the continuity of capillary pressure and wetting-phase flux are met. Looking closely at the saturation and specific interfacial area profiles in the left subdomain, a small kink directly at the interface can be observed. This kink is most probably due to the fact that the determination of the saturation and specific interfacial area jump at the interface is based on the numerical reconstruction of the capillary pressure–saturation curves at the interface as given by Eq.s (5.32) and (5.33). In order to avoid this problem we are planning to extend the interface condition by the discretized version of Eq. (5.27) for the left and right subdomain in the future.

In this section, a redistribution problem posed by Philip [1991] in the context of the classical approach was solved semi-analytically and numerically. This was done in order to validate the numerical model for two-phase flow including interfacial area. An experimental investigation technique was presented.

5.3 Kinetic interphase mass transfer

Most often, when there are two or more fluid phases involved, mass transfer between phases takes place and consequently, the phase compositions change. For some applications, this interphase mass transfer plays a crucial role. This is the case e.g. for DNAPL (dense non-aqueous phase liquid) pool dissolution, for CO_2 storage in geological formations, or for groundwater remediation techniques such as air sparging.

It is common practice to assume a first-order rate of kinetic mass transfer between fluid phases in a porous medium on a macroscopic, i.e. volume-averaged scale which can be expressed as (see e.g. Mayer and Hassanizadeh [2005])

$$Q^\kappa_{\alpha\to\beta} = k^\kappa_{\alpha\to\beta} a_{\alpha\beta} (C^{\prime\kappa}_{\beta,s} - C^\kappa_\beta), \tag{5.58}$$

where $Q^\kappa_{\alpha\to\beta}$ $\left[\frac{kg}{m^3 s}\right]$ is the interphase mass transfer rate of component κ from phase α to phase β, $k^\kappa_{\alpha\to\beta}$ $\left[\frac{m}{s}\right]$ is the mass transfer rate coefficient, $a_{\alpha\beta}$ $\left[\frac{1}{m}\right]$ is the specific interfacial area, i.e. the interfacial area per representative elementary volume (REV), separating phases α and β, $C^{\prime\kappa}_{\beta,s}$ $\left[\frac{kg}{m^3}\right]$ is the solubility limit of component κ in phase β, and finally, C^κ_β $\left[\frac{kg}{m^3}\right]$ is the actual concentration of component κ in phase β. The actual concentration cannot be larger than the solubility limit, $C^\kappa_\beta \leq C^\kappa_{\beta,s}$. The case

$C_\beta^\kappa = C_{\beta,s}^\kappa$ corresponds to the case of local chemical equilibrium. In current models, the interfacial area $a_{\alpha\beta}$ is an unknown parameter. Therefore, current models need to find a way to estimate or get rid of interfacial area. Three different approaches are commonly used,

1. lumping interfacial area into an effective rate coefficient $K_{\alpha \to \beta}^\kappa$ $\left[\frac{1}{s}\right]$ and then empirical estimation of the effective coefficient from a modified Sherwood number (usually done for DNAPL pool dissolution, see e.g. Miller et al. [1990], Powers et al. [1992, 1994], Imhoff et al. [1994], Zhang and Schwartz [2000]),

2. assumption of local equilibrium (e.g. Coats [1980], Young [1984], Allen [1985], Baehr and Corapcioglu [1984], Abriola and Pinder [1985a,b], Parker et al. [1987]), or

3. a dual domain approach (Falta [2000, 2003], van Antwerp et al. [2008]).

Based on thermodynamic considerations, Hassanizadeh and Gray [1980, 1990, 1993a,b] have proposed an alternative, thermodynamically consistent, theory for two-phase flow in porous media. A main characteristic of the alternative model is that it includes not only balance equations for bulk phases, but additionally it contains balance equations for interfaces. The capillary pressure–saturation relationship used in classical models is known to be non-unique; it is dependent on the history of the two-phase flow process. The alternative theory suggests that this non-uniqueness is due to the absence of specific interfacial area in this relationship and it proposes dependence of capillary pressure on saturation *and* specific interfacial area. This dependence has been shown both experimentally (Brusseau et al. [1997], Chen and Kibbey [2006], Culligan et al. [2004], Schaefer et al. [2000], Wildenschild et al. [2002]) and by pore-scale network modeling (Reeves and Celia [1996], Held and Celia [2001], Joekar-Niasar et al. [2008]). The alternative model was recently used by Niessner and Hassanizadeh [2008] who simulated the flow of two immiscible single-component fluids (i.e. without interphase mass transfer) in a porous medium. In this section, we will extend the approach of Niessner and Hassanizadeh [2008] (which was discussed in Sec. 5.1) to model solute transport in two-phase flow including interphase mass transfer. This extension is based on the considerations in Sec. 3.4 and published in Niessner and Hassanizadeh [2009a,b]. To the best of our knowledge, this is the first work where kinetic interphase mass transfer is modeled in the framework of a thermodynamically consistent set of equations involving interfacial dynamics. The extended model allows for mass transfer between the two fluids to be affected by that dynamics. Thus, by making use of the explicit calculation of specific interfacial area, it accounts for kinetic interphase mass transfer in a physically-based way.

It is true that in some current models, specific interfacial area is lumped into a fitting parameter (Sherwood number approach) or used for fitting itself (dual domain approach). But then, this fitting parameter is assumed to be either a constant or a function of saturation. However, we know that interfacial area and saturation are independent variables; dynamics of interfaces are different from dynamics of saturation. So, in a physically-based model of dissolution, one must explicitly calculate interfacial area and its variation, and make the mass transfer rate directly dependent on the interfacial area. Fortunately, this has become also practically possible due to recent advances in the development of techniques for the measurement of interfacial areas. This is a growing field of research.

5.3.1 Classical approaches

In the absence of a physical estimate of fluid–fluid interfacial area, current models of two-phase flow in porous media need to use empirical models or make strong assumptions in order to describe mass transfer processes between fluid phases. To the best of our knowledge, three different approaches have been applied so far: a fitting procedure related to a modified Sherwood number, assumption of local chemical equilibrium, and a dual domain approach. We will briefly present and discuss these approaches in connection with DNAPL pool dissolution and gas–water systems in Sec. 5.3.1.1 and 5.3.1.2.

5.3.1.1 DNAPL pool dissolution

When an immobile pool of DNAPL which is present at residual saturation dissolves into the surrounding groundwater, the kinetics of this mass transfer process play an important role: the dissolution of DNAPL is a rate limited process. In this relatively simple case, only the mass transfer of a DNAPL component from the DNAPL phase into the water phase has to be considered. Without a physically-based estimate of specific interfacial area, the mass transfer coefficient and the specific interfacial area are often lumped into one single parameter k (Miller et al. [1990], Powers et al. [1992,

1994], Imhoff et al. [1994], Zhang and Schwartz [2000]). This yields

$$Q = k(C_s - C). \tag{5.59}$$

Here, C_s is the solubility limit of the DNAPL component in water and C is its actual concentration. The mass transfer coefficient k $\left[\frac{1}{s}\right]$ is related to a modified Sherwood number Sh as

$$k = Sh\frac{D_m}{d_{50}^2}, \tag{5.60}$$

where D_m $\left[\frac{m^2}{s}\right]$ is the aqueous phase molecular diffusion coefficient, and d_{50} [m] is the mean size of the grains. The Sherwood number is then related to Reynolds number Re and DNAPL saturation S_n $[-]$ by

$$Sh = \alpha Re^\beta S_n^\gamma, \tag{5.61}$$

where α, β, and γ are dimensionless fitting parameters.

This approach is empirical and three fitting parameters are required. Although interphase mass transfer is proportional to specific interfacial area, this dependence cannot explicitly be accounted for as the magnitude of specific interfacial area is not known.

An alternative approach for DNAPL pool dissolution has been developed by Falta [2003] who modeled the dissolution of DNAPL components from the DNAPL phase to the water phase by a dual domain approach for a case with simple geometry. For this purpose, they divided the contaminated porous medium into two parts: one that contains DNAPL pools, and one without DNAPL. For their simple case, the dual domain approach combined with an analytical solution for steady-state advection and dispersion provided a means for modeling rate-limited interphase mass transfer. While this approach provided good results for this case with simplified geometry, it might be oversimplified for the modeling of realistic situations.

5.3.1.2 Gas–water systems

For gas–water systems, the situation is more complex than for DNAPL pool dissolution. While for DNAPL pool dissolution one of the phases is stagnant, for gas–water systems, a full two-phase system including two mobile phases has to be considered.

Almost all numerical models are based on the assumption of local chemical equilibrium between phases, most often without any justification. Some models, though, include kinetics of mass transfer by use of a dual domain approach and interpret specific interfacial area as a fitting parameter. We will disuss both types of models in the following.

Local equilibrium models Local equilibrium models for compositional multiphase systems have been introduced and developed by Miller et al. [1990], Powers et al. [1992, 1994], Imhoff et al. [1994], Zhang and Schwartz [2000] and have been used and advanced ever since. The assumption that the composition of a phase is at or close to equilibrium might be good if the time scale of mass transfer is small compared to the time scale of flow. However, if large flow velocities occur as e.g. during air sparging or if the mass transfer process is slow, the assumption of local chemical equilibrium gives completely wrong results, see Falta [2000, 2003] and van Antwerp et al. [2008].

For a gas–water system made up of the components water and air, Henry's Law is employed to determine the mole fraction of air in the water phase, while the mole fraction of water in the gas phase is determined by assuming that water pressure in the gas phase is equal to the saturation vapor pressure. Denoting the water component by superscript w and the air component by superscript a, this yields

$$x_w^a = p_n^a \cdot H_{w-n}^a \qquad (5.62)$$

$$x_n^w = \frac{p_{sat}^w}{p_n}, \qquad (5.63)$$

where x_w^a [−] is the mole fraction of air in the water phase, x_n^w [−] is the mole fraction of water in the gaseous phase, H_{w-n}^a $\left[\frac{1}{Pa}\right]$ is the Henry constant for the dissolution of air in the water phase, p_{sat}^w [Pa] is the saturation vapor pressure of water, p_n^a [Pa] is the partial pressure of air in the gas phase while p_n [Pa] is the gas pressure. The remaining mole fractions result simply from the conditions that mole fractions in each phase have to sum up to one,

$$x_w^w = 1 - x_w^a \qquad (5.64)$$

$$x_n^a = 1 - x_n^w. \qquad (5.65)$$

Dual domain approach Falta [2000] developed a dual domain approach which allows to model kinetic mass transfer during air sparging on the laboratory scale. van Antwerp et al. [2008] showed that the approach of Falta [2000] can also be used to model field-scale experiments.

The basic idea of their dual domain approach is to split each gridblock into two parts: one containing the smaller pores leading to high capillary pressure values and to a local water saturation close to one, and a second subdomain consisting of the larger pores leading to lower capillary pressure values. These larger pores form pathways for the gas during sparging, thus leading to higher gas saturations in those regions. The two domains are coupled through a first-order mass transfer rate as given in Eq. (5.58). This approach allows to estimate the product of the mass transfer coefficient with specific interfacial area as

$$k^\kappa_{\alpha \to \beta} a_{\alpha\beta} = \left(\phi S_g \tau_g D_g + \phi S_w \tau_w D_w / H^n_{w-g}\right)_{I-II} \frac{a_{I-II}}{d_{I-II}}, \quad (5.66)$$

where ϕ [−] is the porosity, S_α [−] is the saturation of phase α, with g denoting the gas phase and w the water phase, τ_α [−] is the phase tortuosity, D_α $\left[\frac{m^2}{s}\right]$ is the phase molecular diffusion coefficient, H^n_{w-g} is a dimensionless Henry coefficient for solution of air in the water phase, a_{I-II} $\left[\frac{1}{m}\right]$ is the specific interfacial area between the coarse-grained subdomain I and the fine-grained subdomain II of a gridblock, and d_{I-II} [m] is the diffusion distance between the two gridblocks. As both specific interfacial area a_{I-II} and diffusion length d_{I-II} are unknown the factor $\frac{a_{I-II}}{d_{I-II}}$ is used as fitting parameter.

5.3.2 Alternative approach including interfaces

In this section, a model for two-phase flow including kinetic interphase mass transfer will be presented that is based on the alternative theory. Therefore, the starting point is the set of equations proposed in Sec. 3.4. In the following, we assume that the composition of the interface does not change which is a reasonable assumption as long as no surfactants are involved. This reduces the number of balance equations to 8, as we can sum up the equations for interface components, both for mass and momentum. Furthermore, we assume that momentum balances can be simplified so far that we end up with extended Darcy equations for both bulk phases and a diffusion-like equation for the interface. We further proceed by applying Fick's law to relate the diffusive fluxes $\underline{j}^\kappa_\alpha$ to known quantities.

The right hand sides of Eq.s (3.104) through (3.106) determine mass transfer. As they still contain micro-scale quantities they need very careful consideration. We consider their advective and diffusive part separately:

For the advective part, we proceed as given in Niessner and Hassanizadeh [2008]. We argue that the advective transfer of mass from the interface to the bulk phases is negligibly small compared to the mass of the bulk phase, but large compared to the mass of the interface. Therefore, the advective mass transfer term $\frac{1}{V} \int_{A_{wn}} [\rho_\alpha X^\kappa_\alpha (\underline{v}_\alpha - \underline{v}_{wn})] \cdot \underline{n}_{wn} \, dA$ is neglected in the bulk phase mass balance equations. But in the interfacial balance equations, we define a general production rate of interfacial area as

$$E_{wn} = \frac{1}{V \cdot \Gamma_{wn}} \int_{A_{wn}} [\rho_w \underline{v}_w (\underline{v}_w - \underline{v}_{wn}) - \rho_n \underline{v}_n (\underline{v}_n - \underline{v}_{wn})] \cdot \underline{n}_{wn} \, dA. \quad (5.67)$$

For the diffusive part of the mass transfer expression in the surface integrals of Eq.s (3.104) through (3.106), we assume that the interphase mass flux can be described by a first-order Fickian expansion. Thus, we have obtained the following formula:

$$\underline{j}^\kappa_\alpha \cdot \underline{n}_{wn} = \pm \frac{\rho_\alpha D^\kappa}{d^\kappa} a_{wn} \left(X^\kappa_{\alpha,s} - X^\kappa_\alpha \right), \quad (5.68)$$

where $D^\kappa \; \left[\frac{m^2}{s}\right]$ is the micro-scale Fickian diffusion coefficient for component κ, d^κ [m] is the diffusion length of component κ, $X^\kappa_{\alpha,s}$ [−] is the solubility limit of component κ in phase α (i.e. the mass fraction corresponding to local chemical equilibrium), and X^κ_α [−] is the micro-scale mass fraction of component κ in phase α at a distance d^κ away from the interface. This is the point in our model where we need to make a strong assumption. We assign a value to the diffusion length corresponding to a pore-scale characteristic length (here assumed to be 10^{-4} m) and assume further, that at this distance away from an interface, we encounter the REV-averaged mass fraction of a component in that phase. Although these estimates are rough, we hope to be able to get more insight into these values and to obtain good estimates for them from pore-scale network modeling in the future. These considerations lead to the following determinate set of equations:

$$\frac{\partial \left(\phi S_w \bar{\rho}_w \bar{X}_w^w\right)}{\partial t} + \nabla \cdot \left(\bar{\rho}_w \bar{X}_w^w \underline{\bar{v}}_w\right) + \nabla \cdot \underline{\bar{j}}_w^w$$
$$= \rho_w Q_w^w - \frac{D^w \bar{\rho}_n}{d^w} a_{wn} \left(X_{n,s}^w - \bar{X}_n^w\right) \quad (5.69)$$

$$\frac{\partial \left(\phi S_w \bar{\rho}_w \bar{X}_w^a\right)}{\partial t} + \nabla \cdot \left(\bar{\rho}_w \bar{X}_w^a \underline{\bar{v}}_w\right) + \nabla \cdot \underline{\bar{j}}_w^a$$
$$= \rho_w Q_w^a + \frac{D^a \bar{\rho}_w}{d^a} a_{wn} \left(X_{w,s}^a - \bar{X}_w^a\right) \quad (5.70)$$

$$\frac{\partial \left(\phi S_n \bar{\rho}_n \bar{X}_n^w\right)}{\partial t} + \nabla \cdot \left(\bar{\rho}_n \bar{X}_n^w \underline{\bar{v}}_n\right) + \nabla \cdot \underline{\bar{j}}_n^w$$
$$= \rho_n Q_n^w + \frac{D^w \bar{\rho}_n}{d^w} a_{wn} \left(X_{n,s}^w - \bar{X}_n^w\right) \quad (5.71)$$

$$\frac{\partial \left(\phi S_n \bar{\rho}_n \bar{X}_n^a\right)}{\partial t} + \nabla \cdot \left(\bar{\rho}_n \bar{X}_n^a \underline{\bar{v}}_n\right) + \nabla \cdot \underline{\bar{j}}_n^a$$
$$= \rho_n Q_n^a - \frac{D^a \bar{\rho}_w}{d^a} a_{wn} \left(X_{w,s}^a - \bar{X}_n^a\right) \quad (5.72)$$

$$\frac{\partial a_{wn}}{\partial t} + \nabla \cdot (a_{wn} \underline{v}_{wn}) = E_{wn} \quad (5.73)$$

$$\underline{\bar{v}}_w = -K \frac{S_w^2}{\mu_w} \left(\nabla p_w - \bar{\rho}_w \underline{g}\right) \quad (5.74)$$

$$\underline{\bar{v}}_n = -K \frac{S_n^2}{\mu_n} \left(\nabla p_n - \bar{\rho}_n \underline{g}\right) \quad (5.75)$$

$$\underline{\bar{v}}_{wn} = -K_{wn} \nabla a_{wn} \quad (5.76)$$

$$p_c = p_n - p_w \quad (5.77)$$

$$S_w + S_n = 1 \quad (5.78)$$

$$\bar{X}_w^w + \bar{X}_w^a = 1 \quad (5.79)$$

$$\bar{X}_n^w + \bar{X}_n^a = 1 \quad (5.80)$$

$$a_{wn} = a_{wn}(S_w, p_c), \quad (5.81)$$

where E_{wn} $\left[\frac{1}{\text{m}\cdot\text{s}}\right]$ is the production rate of specific interfacial area which, based on physical argument, has been modeled by Niessner and Hassanizadeh [2008] as $E_{wn} = -e_{wn}(S_w, p_c) \cdot \frac{\partial S_w}{\partial t}$. Here, the value of e_{wn} $\left[\frac{1}{\text{m}}\right]$ can be obtained by interpolation between two points where its value is known, see Sec. 5.1.1.2. The parameter

K [m²] denotes intrinsic permeability, μ_α [Pa · s] is the dynamic viscosity of phase α, p_n and p_w are the pressures of phase n and phase w whereas p_c is capillary pressure. Furthermore, K_{wn} $\left[\frac{m^3}{s}\right]$ is interfacial permeability. Eq. (5.76) is derived from the basic physical principle that the driving force for movement of interfaces is a gradient in Gibbs free energy. Hassanizadeh and Gray [1990] exploited the second law of thermodynamics leading to a simpler equation which includes gradients in specific interfacial area, in macroscopic interfacial tension, and in saturation as well as gravity as the driving forces. A list of simplifications made by Niessner and Hassanizadeh [2008] leaves us with specific interfacial area as the only driving force and the formula of Eq. (5.76) where interfacial permeability is the product of the inverse of resistance to flow of interfaces and macroscopic interfacial tension.

The macro-scale mass fluxes $\bar{j}_\alpha^\kappa$ are calculated from a Fickian dispersion equation by

$$\bar{j}_\alpha^\kappa = -\bar{\rho}_\alpha \bar{D}_\alpha^\kappa \nabla \bar{X}_\alpha^\kappa, \qquad (5.82)$$

where $\bar{D}_\alpha^\kappa$ is the macro-scale dispersion coefficient of component κ in phase α. Note that in Eq.s (5.69) through (5.72) we have acknowledged the fact that a source of a component in one of the phases must correspond to a sink of that component of equal magnitude in the other phase. The relationship $a_{wn}(S_w, p_c)$ can be obtained from experiments or from pore-scale network models.

5.3.3 Numerical test case

Through a hypothetical test case, we show how the alternative model for two-phase flow in porous media performs numerically and compares to current approaches, specifically to models which neglect kinetics of interphase mass transfer and assume local chemical equilibrium. We are aware of the fact that a comparison of a kinetic model to an equilibrium model might seem inappropriate; we have chosen this way of comparison, though, as currently, equilibrium models are generally used for two-phase systems where both fluid phases are flowing, such as for CO_2 storage or for degassing processes in fractures due to hydrogen escaping from atomic waste disposal sites. This choice is made although kinetics are definitely important in a number of cases. So this numerical study is chosen in the spirit of comparing what is currently done in case of two flowing fluid phases to what is made possible by the presented physically based kinetic approach.

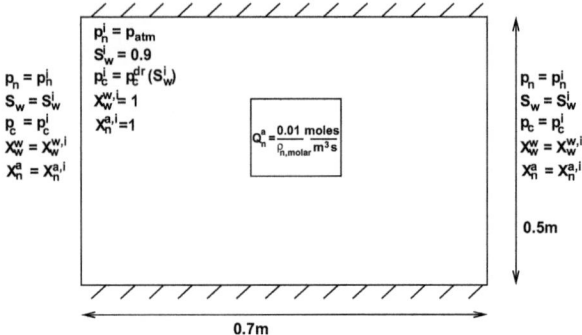

Figure 5.15: Setup of the test example.

We consider a simple model setup where kinetic mass transfer processes will definitely occur: dry gas is injected into a domain which contains mostly water. We assume that we are at the start of a main drainage process. Both water and gas phase, also inside the domain, are initially pure: the water phase only consists of water component and the gas phase is devoid of water. We are aware of the fact that this setup is probably not relevant and hard to realize experimentally. We have chosen it, though, because kinetic mass transfer will here be of importance throughout the domain. In the following, we will first describe the setup of the numerical example in more detail. Next, we will discuss choices of crucial parameters and parameter functions, and finally, we will show simulation results comparing the new model and the current (equilibrium) model.

Setup of the numerical example The detailed setup of this example is shown in Fig. 5.15. Initial values and boundary values at the left and right boundaries are identical (superscript i indicates initial values) while the top and bottom boundaries are closed. At the beginning, the domain contains pure water ($\bar{X}_w^{i,w} = 1$) with $S_w^i = 0.9$, and also the gas phase is pure with $\bar{X}_n^{i,a} = 1$. Initial pressures are chosen in a way that gas pressure is at atmospheric pressure and capillary pressure is at the value corresponding to primary drainage at a saturation of S_w^i. In the middle of the domain, pure (dry) gas is injected at a constant rate of $Q_n^a = \frac{0.1}{\rho_{n,molar}} \frac{\text{moles}}{\text{m}^3 \cdot \text{s}}$, where $\rho_{n,molar}$ is the molar density of the gas phase (depending on gas pressure).

Parameters For the example at hand, we choose fluid parameters (densities, viscosities, and diffusion constants) of water and air, for the matrix parameters we take typical soil parameters, here $K = 10^{-10}$ m² and $\phi = 0.3$. Analysis of experimental studies by Crandall et al. [2008] following the procedure of Niessner et al. [2008] revealed that interfacial permeability is very small, here we take a rough estimate of $K_{wn} = 10^{-5} \frac{m^3}{s}$. Solubility limits are calculated as in the equilibrium model, i.e. mole fractions are calculated from the local equilbrium assumption as given in Eq.s (5.62) and (5.63) with $H^a_{w-n} = 1.5 \cdot 10^{-10} \left[\frac{1}{Pa}\right]$ and $p^w_{sat} = 2339.2$ Pa and transformed to mass fractions.

A constitutive function needs to be given for the interfacial area–capillary pressure–saturation surface $a_{wn} = a_{wn}(S_w, p_c)$. For our numerical simulation, we use the following formula suggested by Joekar-Niasar et al. [2008] based on data from a pore network model:

$$a_{wn}(S_w, p_c) = -313.6\,\frac{1}{m} + 5535\,\frac{1}{m}S_w + 0.085\,\frac{1}{Pa\cdot m}p_c - 3937\,\frac{1}{m}S_w^2$$
$$- 0.307\,\frac{1}{Pa\cdot m}S_w \cdot p_c - 5.2\cdot 10^{-6}\,\frac{1}{Pa^2 m}p_c^2. \quad (5.83)$$

Another crucial issue is how to describe both macro-scale and micro-scale diffusion and dispersion. As the molecular diffusion coefficient in a gas phase is several orders of magnitude larger than in the liquid phase, we only consider the macro-scale diffusion of components in the gas phase. The macro-scale diffusive fluxes for a binary mixture are given by

$$\bar{D}^\kappa_n = \phi \tau S_n D^\kappa, \quad (5.84)$$

where τ is the tortuosity which can be obtained from the model of Millington and Quirk [1961] as

$$\tau = \phi^{\frac{1}{3}} S_n^{\frac{7}{3}}. \quad (5.85)$$

Results Fig. 5.16 shows a comparison of the results obtained using the alternative model denoted here by 2p2cia (2-phase–2-component model with interfacial area) and a current (classical) model which assumes local equilibrium (2p2c—2-phase–2-component model). For that purpose, we compare the contour lines of S_w, p_n, $\bar{X}^w_w$, $\bar{X}^a_n$ after 0, 5s, 75s, and 150s. Additionally, we show contour lines of p_c and a_{wn} of the 2p2cia model.

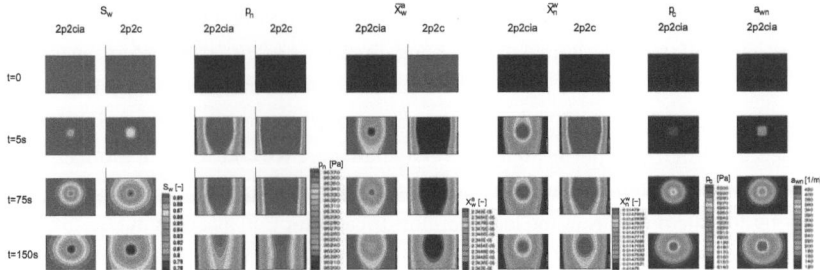

Figure 5.16: Contour plots of S_w, p_n, $\bar{X}_w^w$, $\bar{X}_n^a$, p_c, a_{wn} at four selected time steps.

It can be seen that the contour lines of S_w and also p_n are qualitatively the same for both models, although the actual values differ slightly. Mass fractions of air in the water phase $\bar{X}_w^a$ and of water in the gas phase $\bar{X}_n^w$ are substantially different. This is due to the fact that kinetic mass transfer is accounted for in the 2p2cia model whilst local equilibrium is assumed to hold in the 2p2c model. A very interesting fact is the shape of these contour lines: while the contour lines for both mass fractions in the 2p2c model qualitatively follow the shape of the pressure contours, the contour lines of mass fractions in the 2p2cia model are a mixture of the contour shapes of specific interfacial area and of pressure. This is due to the fact that the local equilibrium values of the mass fractions are determined solely by the pressure field, while the kinetic part of interphase mass transfer is proportional to specific interfacial area.

Considering two-phase systems like the one discussed above, the question arises whether for the particular system of interest kinetic interphase mass transfer is important or whether the assumption of local chemical equilibrium is justified. To answer this question, Niessner and Hassanizadeh [2009a] non-dimensionalized the equation system given by Eq.s (5.69) through (5.81) and identified relevant dimensionless numbers, see Sec. 6.2.2.

5.4 Kinetic interphase energy transfer

5.4.1 Background

Modeling kinetic interphase mass and / or heat transfer in two-phase flow in porous media is highly challenging: because mass and energy are transfered across phase-interfaces, they are highly dependent on interfacial areas, at least, as long as local chemical and thermal equilibrium is not reached. Unfortunately, interfacial areas do not appear as variables in classical two-phase flow models. Therefore, these classical models either are only applicable to equilibrium situations with respect to mass and energy transfer, or have to employ some empirical transfer relationship, where a kinetic transfer coefficient is obtained by fitting to experimental data. In conclusion, there is a strong need for physically motivated models for two-phase flow that include interfacial areas, thus allowing to properly account for non-equilibrium interphase mass and heat transfer. Fig. 5.17 illustrates this situation: on the pore

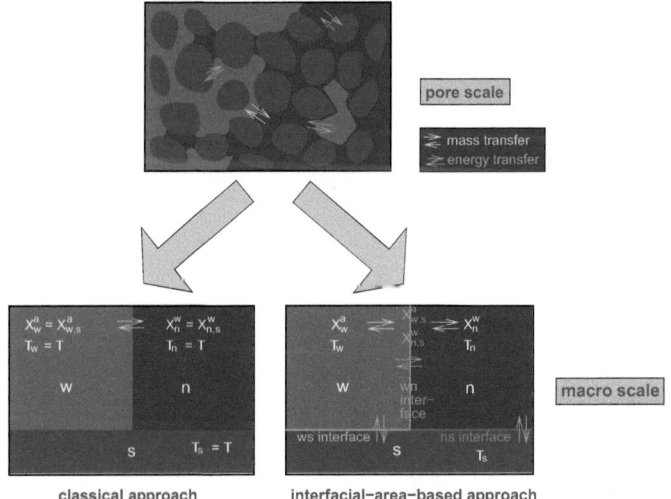

Figure 5.17: Conceptual models with respect to heat and energy transfer on the pore and macro scale using a classical approach and using an interfacial-area-based model. The wetting fluid phase is denoted by w, the non-wetting fluid phase by n, and the solid phase by s.

scale, mass is transferred across the fluid–fluid interfaces and energy is transferred between all three phases. Phases are denoted by α and may be the wetting fluid phase (index w), the non-wetting fluid phase (index n), or the solid phase (index s). Temperature and composition of phases may vary continuously throughout the domain. The classical procedure is to average this pore-scale situation over a representative elementary volume (REV) yielding porosity, phase saturations and one average temperature T [K]. Also, if empirical models are to be avoided, local chemical and thermal equilibrium between phases with respect to mass transfer has to be assumed. This means that equilibrium relationships, such as Henry's law and Raoult's law, are used. Thus, mass fractions for component κ in phase α, X_α^κ [-], that correspond to the equilibrium values $X^\kappa_{\alpha,s}$ [-] are determined. A model using interfacial areas could yield mass fractions within the phases that do not correspond to the equilibrium values, or allow for different temperatures T_α for different phases. Such a model has been developed by Niessner and Hassanizadeh [2009c] and will be presented in this section.

The basis for the following considerations is the macro-scale set of energy balance equations derived in Sec. 3.5. For the treatment and detailed discussion of mass and momentum balances we refer to the work of Niessner and Hassanizadeh [2009a,b], see Sec. 5.3.2. An upscaling of the complete set of balance equations leads to the following set of macro-scale conservation equations for mass of the two components (called w- and a) in w-phase and n-phase, for mass of wn-interface, momentum of

w-phase, n-phase, wn-interface, and energy of w-, n-, and s-phase:

$$\frac{\partial (\phi S_w \rho_w X_w^w)}{\partial t} + \nabla \cdot (\rho_w X_w^w \underline{v}_w) + \nabla \cdot \left(\underline{j}_w^w\right) = \rho_{w,Q} Q_w^w - \frac{D^w \rho_n}{d^w} a_{wn} \left(X_{n,s}^w - X_n^w\right) \quad (5.86)$$

$$\frac{\partial (\phi S_w \rho_w X_w^a)}{\partial t} + \nabla \cdot (\rho_w X_w^a \underline{v}_w) + \nabla \cdot \left(\underline{j}_w^a\right) = \rho_{w,Q} Q_w^a + \frac{D^a \rho_w}{d^a} a_{wn} \left(X_{w,s}^a - X_w^a\right) \quad (5.87)$$

$$\frac{\partial (\phi S_n \rho_n X_n^w)}{\partial t} + \nabla \cdot (\rho_n X_n^w \underline{v}_n) + \nabla \cdot \left(\underline{j}_n^w\right) = \rho_{n,Q} Q_n^w + \frac{D^w \rho_n}{d^w} a_{wn} \left(X_{n,s}^w - X_n^w\right) \quad (5.88)$$

$$\frac{\partial (\phi S_n \rho_n X_n^a)}{\partial t} + \nabla \cdot (\rho_n X_n^a \underline{v}_n) + \nabla \cdot \left(\underline{j}_n^a\right) = \rho_{n,Q} Q_n^a - \frac{D^a \rho_w}{d^a} a_{wn} \left(X_{w,s}^a - X_w^a\right) \quad (5.89)$$

$$\frac{\partial a_{wn}}{\partial t} + \nabla \cdot (a_{wn} \underline{v}_{wn}) = E_{wn} \quad (5.90)$$

$$\frac{\partial (\phi \rho_w S_w h_w)}{\partial t} + \nabla \cdot (\phi \rho_w S_w h_w \underline{v}_w) + 0.25 C_{p,wn} (T_n - T_w) \Gamma_{wn} E_{wn}$$
$$= \nabla \cdot \left(\underline{\underline{D}}_w^{th} \nabla T_w\right) + \frac{a_{wn} \lambda_{wn}}{d_{wn}^T} (T_n - T_w)$$
$$+ \frac{a_{ws} \lambda_{ws}}{d_{ws}^T} (T_s - T_w) + \rho_{w,Q} h_{w,Q} \left(Q_w^w + Q_w^a\right) \quad (5.91)$$

$$\frac{\partial (\phi \rho_n S_n h_n)}{\partial t} + \nabla \cdot (\phi \rho_n S_n h_n \underline{v}_n) + 0.25 C_{p,wn} (T_w - T_n) \Gamma_{wn} E_{wn}$$
$$= \nabla \cdot \left(\underline{\underline{D}}_n^{th} \nabla T_n\right) - \frac{a_{wn} \lambda_{wn}}{d_{wn}^T} (T_n - T_w)$$
$$+ \frac{a_{ns} \lambda_{ns}}{d_{ns}^T} (T_s - T_n) + \rho_{n,Q} h_{n,Q} \left(Q_n^w + Q_n^a\right) \quad (5.92)$$

$$\rho_s C_{p,s} \frac{\partial (1-\phi) T_s}{\partial t} = \nabla \cdot \left(\underline{\underline{D}}_s^{th} \nabla T_s\right) - \frac{a_{ws} \lambda_{ws}}{d_{ws}^T} (T_s - T_w)$$
$$- \frac{a_{ns} \lambda_{ns}}{d_{ns}^T} (T_s - T_n) \quad (5.93)$$

$$\underline{v}_w = -\underline{\underline{K}} \frac{S_w^2}{\mu_w} \left(\nabla p_w - \rho_w \underline{g}\right) \quad (5.94)$$

$$\underline{v}_n = -\underline{\underline{K}} \frac{S_n^2}{\mu_n} \left(\nabla p_n - \rho_n \underline{g}\right) \quad (5.95)$$

$$\underline{v}_{wn} = -\underline{\underline{K}}_{wn} \nabla a_{wn} \quad (5.96)$$

This system of balance equations is supplemented by the following additional relations:

$$\underline{j}_w^w = -D_w^w \nabla (\rho_w X_w^w) \tag{5.97}$$

$$\underline{j}_w^a = -D_w^a \nabla (\rho_w X_w^a) \tag{5.98}$$

$$\underline{j}_n^w = -D_n^w \nabla (\rho_n X_n^w) \tag{5.99}$$

$$\underline{j}_n^a = -D_n^a \nabla (\rho_n X_n^a) \tag{5.100}$$

$$\underline{\underline{D}}_w^{th} = \lambda_w \underline{\underline{I}} + \alpha_T \rho_w C_{p_w} |\underline{v}_w| \underline{\underline{I}} + (\alpha_L - \alpha_T) \rho_w C_{p_w} \frac{\underline{v}_w \otimes \underline{v}_w}{|\underline{v}_w|} \tag{5.101}$$

$$\underline{\underline{D}}_n^{th} = \lambda_n \underline{\underline{I}} + \alpha_T \rho_n C_{p_n} |\underline{v}_n| \underline{\underline{I}} + (\alpha_L - \alpha_T) \rho_n C_{p_n} \frac{\underline{v}_n \otimes \underline{v}_n}{|\underline{v}_n|} \tag{5.102}$$

$$\underline{\underline{D}}_s^{th} = \lambda_s \underline{\underline{I}} \tag{5.103}$$

$$p_c = p_n - p_w \tag{5.104}$$

$$S_w + S_n = 1 \tag{5.105}$$

$$X_w^w + X_w^a = 1 \tag{5.106}$$

$$X_n^w + X_n^a = 1 \tag{5.107}$$

$$a_{wn} = a_{wn}(S_w, p_c) \tag{5.108}$$

$$E_{wn} = -e_{wn}(S_w, p_c) \frac{\partial S_w}{\partial t}, \tag{5.109}$$

The overbars and averaging brackets have been omitted as we only have macro-scale variables in these equations. The variable ϕ [-] denotes porosity; j_α^κ $\left[\frac{\text{kg} \cdot \text{m}^2}{\text{s}}\right]$ is a dispersive flux of component κ in phase α. The external sink or source Q_α^κ $\left[\frac{1}{\text{s}}\right]$ is multiplied by a source density $\rho_{\alpha,Q}$ that may be different from the system phase density in case of an external source. The micro-scale diffusion coefficient of component κ is given by D^κ $\left[\frac{\text{m}^2}{\text{s}}\right]$ and the diffusion length by d^κ [m]; $\underline{v}_{wn}$ $\left[\frac{\text{m}}{\text{s}}\right]$ is interfacial velocity. The enthalpy $h_{\alpha,Q}$ is the α-phase enthalpy related to the sink / source (which might be different from the phase enthalpy in case of an external source). The intrinsic permeability tensor is denoted by $\underline{\underline{K}}$ [m^2] and interfacial permeability tensor by $\underline{\underline{K}}_{wn}$ $\left[\frac{\text{m}^3}{\text{s}}\right]$; μ_α [Pa · s] is dynamic viscosity and p_α [Pa] the pressure of phase α, $\underline{g}$ $\left[\frac{\text{m}}{\text{s}^2}\right]$ is the gravity vector and p_c is the capillary pressure (the pressure difference between non-wetting and wetting phase pressure). Finally, D_α^κ $\left[\frac{\text{m}^2}{\text{s}}\right]$ is the macro-scale diffusion coefficient of component κ in phase α.

Porosity ϕ and solid density ρ_s are usually assumed to be constant; fluid phase densities ρ_w and ρ_n are either assumed to be constant or given by equations of state as a function of phase pressures and temperatures. The production rate coefficient of specific interfacial area, e_{wn}, and specific interfacial areas $a_{\alpha\beta}$ are given as functions of wetting-phase saturation S_w and capillary pressure p_c, see Eq.s (5.108) and (5.109). The enthalpy of a liquid phase is a function of temperature and the phase pressure while the enthalpy of a gaseous phase additionally depends on its composition; e.g. for a gas–water system, one needs to specify $h_w = h_w(T, p_w)$ and $h_n = h_n(T, p_n, X_n^a, X_n^w)$. Densities and enthalpies of external sinks and sources are denoted by $\rho_{\alpha,Q}$ and $h_{\alpha,Q}$. They are equal to the internal value in case of a sink, $\rho_{\alpha,Q} = \rho_\alpha$ and $h_{\alpha,Q} = h_\alpha$. In case of a source, $\rho_{\alpha,Q}$ and $h_{\alpha,Q}$ are not related to the internal values of density or enthalpy and must be specified independently.

Macro-scale diffusion is much larger in gaseous phases than in liquid phases. Therefore, it is often neglected for the liquid phases, and for the gaseous phase, it is estimated by

$$D_n^\kappa := \bar{D}_n^\kappa = \phi \tau S_n D^\kappa, \tag{5.110}$$

where τ [-] is the tortuosity which can be obtained from the model of Millington and Quirk [1961] as

$$\tau = \phi^{\frac{1}{3}} S_n^{\frac{7}{3}}. \tag{5.111}$$

In these equations, one may identify terms such as $\frac{D^w \rho_n a_{wn}}{d^w}$ and $\frac{D^a \rho_w a_{wn}}{d^a}$ as effective mass transfer coefficients. Similarly, there are effective heat transfer coefficients with the general form of $\frac{a_{\alpha\beta} \lambda_{\alpha\beta}}{d_{\alpha\beta}^T}$. The advantage of this approach is that the transfer coefficients depend directly on specific interfacial area and they change as the amount of interfaces in the system changes.

5.4.2 First numerical example: drying of a porous medium

In order to illustrate the role of non-equilibrium effects with respect to both interphase mass and energy transfer, we consider a typical problem of drying of a porous medium as may occur e.g. in the food, paper, and textile industry. Dry hot air is injected in order to dry an initially wet porous medium. We consider two phases: the non-wetting phase (or gaseous) phase and the wetting phase (a liquid). Each phase is assumed to consist of two components. Because we are only interested to show the principal effects, we assume these two components to be air and water in a typical sand medium as parameters are easily available. A detailed list of all parameter values assigned in the numerical model, both "classical" and "interfacial-area-based" parameter values, is given in Tab.s 5.2 and 5.3.

We consider a setup as shown in Fig. 5.18. This setup consists of a horizontal domain that is closed along the sides (top and bottom in the figure) and that is subjected to a gradient in non-wetting phase pressure from left to right in the undisturbed situation. This system is assumed to be at the following initial conditions: a temperature of 293 K, non-wetting phase at atmospheric pressure, water saturation of 0.9, a corresponding capillary pressure based on the primary drainage curve, and mass fractions that correspond to the local equilibrium conditions as prescribed by

Parameter	K	ϕ	ρ_w	ρ_n	ρ_s	μ_w	μ_n	D^w	D^a
Unit	$[\text{m}^2]$	$[-]$	$\left[\frac{\text{kg}}{\text{m}^3}\right]$	$\left[\frac{\text{kg}}{\text{m}^3}\right]$	$\left[\frac{\text{kg}}{\text{m}^3}\right]$	$[\text{Pa}\cdot\text{s}]$	$[\text{Pa}\cdot\text{s}]$	$\left[\frac{\text{m}^2}{\text{s}}\right]$	$\left[\frac{\text{m}^2}{\text{s}}\right]$
Value	$3\cdot 10^{-11}$	0.3	998	1.293	1602	10^{-3}	$1.7\cdot 10^{-5}$	10^{-5}	10^{-9}

Parameter	λ_w	λ_n	λ_s	$C_{p,w}$	$C_{p,n}$	$C_{p,s}$	α_L	α_T
Unit	$\left[\frac{\text{W}}{\text{m}\cdot\text{K}}\right]$	$\left[\frac{\text{W}}{\text{m}\cdot\text{K}}\right]$	$\left[\frac{\text{W}}{\text{m}\cdot\text{K}}\right]$	$\left[\frac{\text{J}}{\text{kg}\cdot\text{K}}\right]$	$\left[\frac{\text{J}}{\text{kg}\cdot\text{K}}\right]$	$\left[\frac{\text{J}}{\text{kg}\cdot\text{K}}\right]$	$[\text{m}]$	$[\text{m}]$
Value	0.6	0.025	0.5	4.18	1.01	0.79	0.01	0.001

Table 5.2: "Classical" parameter values for the numerical example.

Parameter	a_{00}	a_{01}	a_{11}	a_{10}	a_{20}	a_{02}	a_s
Unit	$\left[\frac{1}{m}\right]$	$\left[\frac{1}{m\cdot Pa}\right]$	$\left[\frac{1}{m\cdot Pa}\right]$	$\left[\frac{1}{m}\right]$	$\left[\frac{1}{m}\right]$	$\left[\frac{1}{m\cdot Pa^2}\right]$	$\left[\frac{1}{m}\right]$
Value	-313.6	0.085	-0.307	5535.487	-3936.8	$-5.23\cdot 10^{-6}$	5000

Parameter	K_{wn}	$E_{wn,\alpha}$	d^κ	$d^T_{\alpha\beta}$
Unit	$\left[\frac{m^3}{s}\right]$	$\left[\frac{1}{m\cdot s}\right]$	$[m]$	$[m]$
Value	10^{-5}	$\frac{1}{2}E_{wn}$	10^{-4}	10^{-2}

Table 5.3: Additional parameters for the numerical example using the interfacial-area-based model approach.

Henry's Law and Raoult's Law. The system is perturbed by injection of dry air at 333 K at a constant rate $Q_n^a = \frac{0.2}{\rho_{n,molar}}\frac{\text{moles}}{\text{m}^3\text{s}}$ in a square-shaped part of the domain (see Fig. 5.18). As constitutive relationship for $a_{wn}(S_w, p_c)$, the surface obtained by Joekar-Niasar et al. [2008] from a pore network model is used, see Eq. (5.83). For the specific solid–fluid interfacial area relationships, $a_{ws}(S_w, p_c)$ and $a_{ns}(S_w, p_c)$, the approximate reconstruction method based on the a_{wn} surface is used which was introduced in Sec. 3.5.

Comparisons between the interfacial-area-based model, which is able to account for kinetic interphase mass and energy transfer (denoted by "new approach") and a classical model, which assumes local equilibrium with respect to both mass and energy transfer (denoted by "classical approach") are shown. Note that in the classical approach it is not possible to explicitly prescribe a source of dry air of a certain temperature in a certain phase. Instead, a source of a-component at a certain temperature can be given without fixing a phase; but then, the a-component will automatically be distributed between the phases according to the equilibrium requirement and the temperatures of all three phases in the domain are the same.

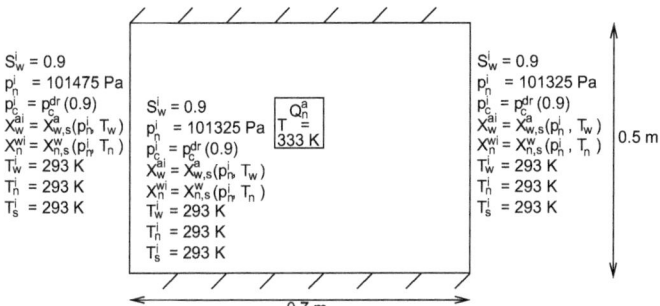

Figure 5.18: Setup of the numerical example.

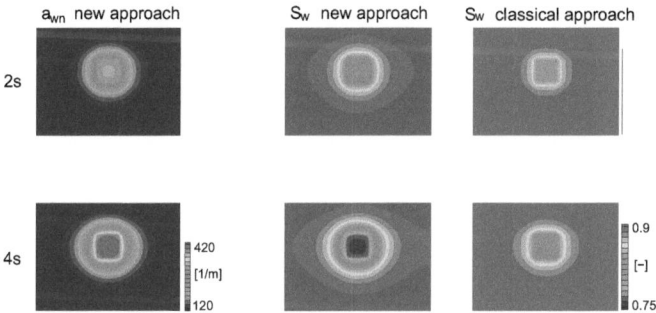

Figure 5.19: Contour lines of a_{wn} for the new approach and S_w for new and classical approach after 2 and 4 s.

Fig.s 5.19 through 5.21 show contour lines of specific interfacial area, saturation, mass fractions, and temperature after 2 s and after 4 s. Fig. 5.19 shows that wetting-phase saturation decreases in the injection region as the porous medium is dried while specific interfacial area increases. Obviously, there is some difference in saturations using the classical and the new approach. This may be due to the fact that the source term is explicitly given as dry air in the gaseous phase (new approach) while a source term of air component is specified in the classical approach. Fig. 5.20 shows that there is a strong non-equilibrium effect with respect to mass transfer as contour lines between equilibrium and kinetic model are completely different. This difference is due to the fact that in the classical approach, the air added to the system will instantaneously redistribute among phases to yield equilibrium composition while in the new approach, the air component is added to the gas phase and needs a non-zero time to be transferred to the water phase to finally yield equilibrium com-

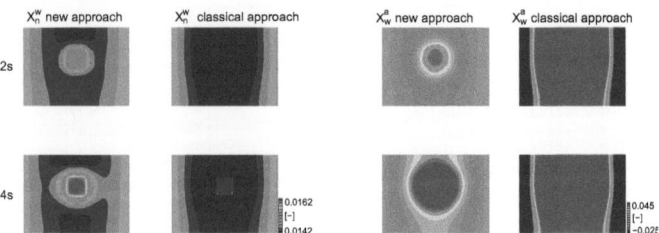

Figure 5.20: Contour lines of X_n^w and X_w^a for new and classical approach after 2 and 4 s.

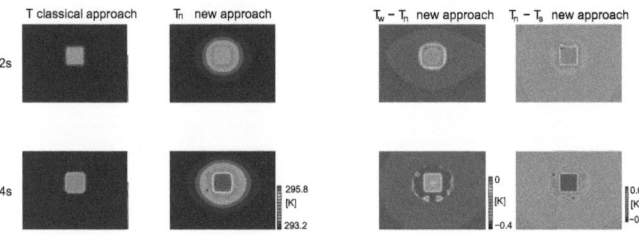

Figure 5.21: Contour lines of T for the classical approach and T_n as well as $T_w - T_n$, $T_n - T_s$ for the new approach after 2 and 4 s.

position. Also, from Fig. 5.21 it becomes obvious that non-equilibrium effects with respect to energy transfer play a significant role during the drying process as $T_w - T_n$ and $T_n - T_s$ have non-zero values. The maximum reported difference in this example was 0.4 K, but the non-equilibrium effect may be more pronounced if porous materials of larger grain sizes are used. In that case, specific interfacial areas and thus also transfer rates may be significantly smaller leading to larger non-equilibrium effects. Another interesting observation from Fig. 5.21 is that the difference in phase temperatures is at a maximum not in the injection region, but in a "ring" around the injection region. This is due to the fact that in the source zone, two different effects compete: on the one hand, air of a higher temperature than the surrounding temperature is injected leading the system away from equilibrium. But on the other hand, specific interfacial areas in that domain increase, which results in higher mass transfer rates and thus, pushes the system towards equilibrium. Obviously, the sum of these effects is such that non-equilibrium is most pronounced not in the source zone, but in a ring around the source zone.

5.4.3 Second numerical example: evaporator

This second numerical example is taken from Ahrenholz et al. [2009] and deals with the processes in an important technical device, an evaporator. Thus, we consider a setup which is relevant in many industrial processes where a product needs to be concentrated (e.g. foods, chemicals, and salvage solvents) or dried through evaporation of water. The aqueous solution containing the desired product is fed into the evaporator mostly consisting of microchannels and then passes a heat source. The heat converts part of the water in the solution into vapor and the vapor is subsequently removed from the solution. We model the heating and evaporation process through a setup as shown in Fig. 5.22. This means, we consider a horizontal do-

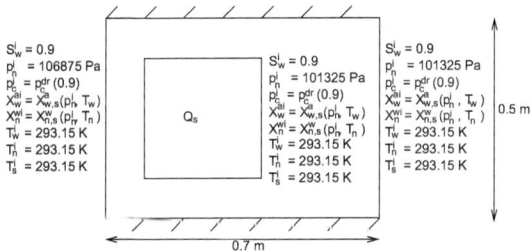

Figure 5.22: Setup of the numerical example: water passes a heat source and is evaporated.

main that is closed along the sides (top and bottom in the figure) and that is subjected to a gradient in wetting phase pressure from left to right in the undisturbed situation. This system is assumed to be at the following initial conditions: a temperature of 293 K, non-wetting phase at atmospheric pressure, water saturation of 0.9, a corresponding capillary pressure based on the primary drainage curve, and mass fractions that correspond to the local chemical equilibrium conditions as prescribed by Henry's Law and Raoult's Law. Constitutive relationships for specific fluid–fluid interfacial areas, $a_{wn}(S_w, p_c)$, but also for specific solid-fluid interfacial areas, $a_{ws}(S_w, p_c)$ and $a_{ns}(S_w, p_c)$ were obtained by Ahrenholz et al. [2009] by Lattice–Boltzmann simulations in a porous medium with a "real" porous medium geometry obtained from a CT scan. Porosity and intrinsic permeability were given consistently for that porous block. The flowing water is evaporated by heating the porous medium with a rate of Q_s in a square-shaped part of the domain (see Fig. 5.22). Note that the heat source heats up only the walls of the microchannels (the solid phase) and the heat is then transferred from the solid phase to the water phase.

For comparison, the same setup is used for simulations using a classical two-phase flow model, where—in the absence of interfacial areas as parameters—local chemical and thermal equilibrium is assumed. This means, that the heat source cannot be defined for the solid phase only; instead, the applied heat will instantaneously lead to a heating of all three phases.

Water saturation and water–gas specific interfacial area is shown in Fig. 5.23 for the situation 17 seconds after the heat source was switched on. The saturation distribution of the interfacial-area-based model is compared to that of the classical model. Obviously, water saturation decreases in the heated region due to the evaporation of water. Downstream of the heated zone, water saturation increases indicating that in the colder regions, water condenses again. Due to the decrease in water saturation, gas–water interfacial areas are created. The classical model predicts a much lower decrease in water saturation in the heated region.

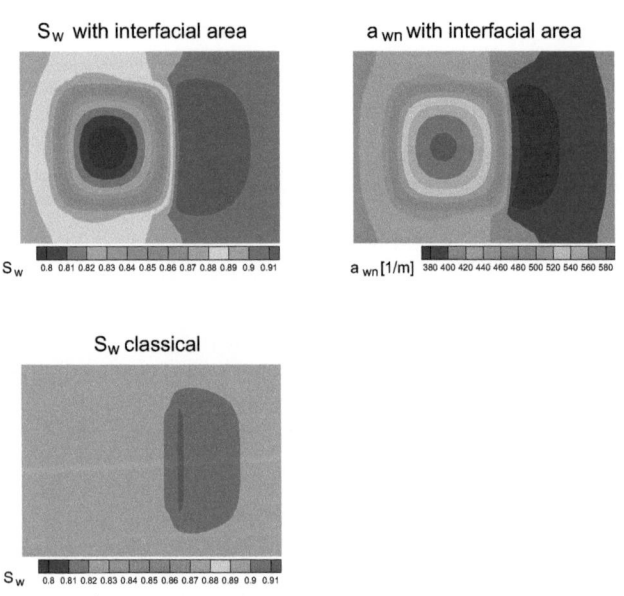

Figure 5.23: Water saturation and water–gas specific interfacial area after 17 s.

Fig. 5.24 shows the mass fractions of air disolved in the water phase after 17 seconds. In the top row, the mass fraction predicted by the interfacial-area-based model is shown as well as the equilibrium value (solubility limit) predicted by the interfacial-area-based model. In the lower left picture, results obtained using the classical model are shown. Clearly, chemical non-equilibrium effects occur, but the classical model predicts approximately the same results as the equilibrium values in the interfacial-area-based model. This is due to the fact that the classical two-phase flow approach always assumes local equilibrium and can only represent mass fractions corresponding to the equilibrium values. In reality, however, interphase mass

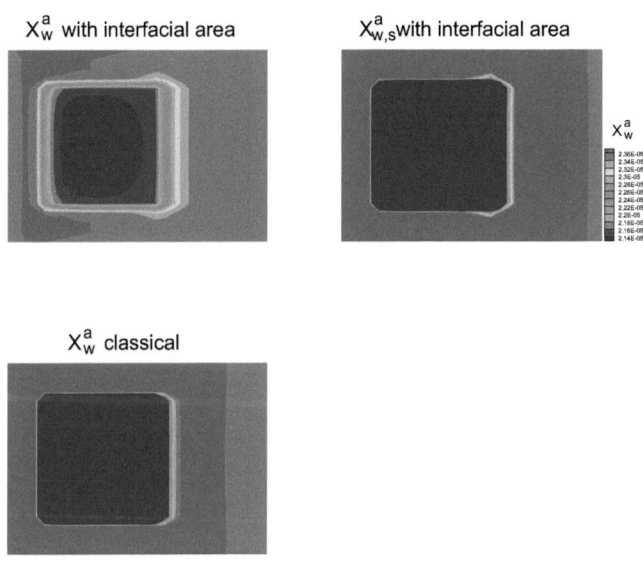

Figure 5.24: Mass fractions of air in water after 17 s. Top left: results using the interfacial area based model, top right: equilibrium values given by the interfacial-area-based model, bottom: results given by the classical model.

transfer is limited by the water–gas interfacial area and thus, equilibrium does not occur instantaneously. Only through an explicit inclusion of interfacial areas, non-equilibrium effects can be accounted for in a physically based way.

The analogous comparison is shown in Fig. 5.25 with respect to the mass fractions of vapor in the gas phase. Here, the mass fractions in the interfacial-area-based model are very close to the equilibrium values, but larger differences to the classical model can be detected.

Fig. 5.26 shows the temperatures of the three phases (water w, gas n, solid s) after 17 s (upper row) using the interfacial-area-based model. In the lower row, the temperature given by the classical model is shown. It can be seen that a lower temperature rise is predicted by the classical model than by the interfacial-area-based model. Fig. 5.27 shows the temperature differences between the phases using the interfacial area based model. Obviously there are differences of up to more than 8 K between the phases. The classical two-phase model assumes local thermal equilibrium, i.e. $T_w = T_n = T_s$. In reality, however, the heat exchange between the phases is

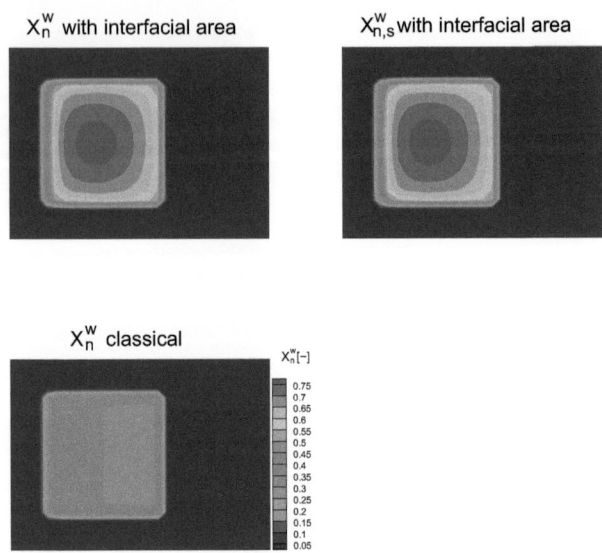

Figure 5.25: Mass fractions of vapor in the gas phase after 17 s. Top left: results using the interfacial area based model, top right: equilibrium values given by the interfacial-area-based model, bottom: results given by the classical model.

restricted by the respective interfacial areas and the temperature equlization does not take place instantaneously. Only by including fluid–fluid and fluid–solid interfacial areas, the kinetic behavior and the resulting non-equilibrium can be described.

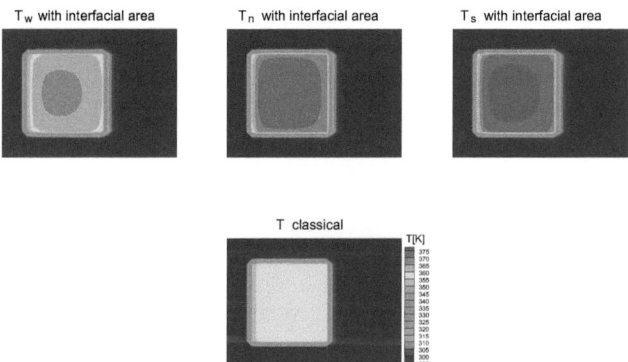

Figure 5.26: Temperatures of the phases after 17 s using the interfacial-area-based model (upper row) and the classical model (lower row).

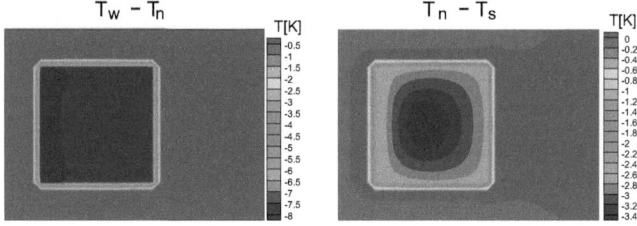

Figure 5.27: Temperature differences between wetting and non-wetting phase (left hand side) and between non-wetting and solid phase (right hand side) after 17 s.

5.4.4 Discussion

These numerical examples are set up to illustrate that by including interfacial areas, kinetic interphase mass and energy transfer processes can be modeled with transfer rates depending on interfacial areas. It could be shown that resulting temperature and concentration fields will be different from those obtained by classical models which assume local chemical and thermal equilibrium. Obviously, parameter values of the new model need to be determined experimentally in order to make quantitative predictions. Of course, some parameters also occur in classical models or are material properties, such as heat capacities or thermal conductivities (see Tab. 5.2). Some others, however, only occur in the interfacial-area-based model. Although some of the "classical" parameters, such as dispersivities, are also not easy to determine, we focus our attention on the parameters additionally needed for the interfacial-area-based model.

The fitting parameters a_{ij} of the $a_{wn}(S_w, p_c)$ surface can be obtained through a fit to e.g. the formula

$$a_{wn}(S_w, p_c) = a_{00} + a_{01}p_c + a_{02}p_c^2 + a_{10}S_w + a_{11}p_cS_w + a_{20}S_w^2 \tag{5.112}$$

to experimental or pore-network data. Fortunately, an increasing amount of data is becoming available as more and more research is being done to determine this relationship through micro tomography, imaging of micro-models, interfacial tracer techniques, or pore-network models. The parameters used here were obtained by fitting this equation to data from a pore-network model of Joekar-Niasar et al. [2008] (drying example) or to data from a multi-phase Lattice-Boltzmann simulation of Ahrenholz et al. [2009] (evaporator example). The specific surface area a_s is also not difficult to obtain; here a typical value for sand was chosen. The interfacial permeability K_{wn} definitely needs to be determined much more carefully from experiments. For now, we have made a rough estimation of its value based on data from flow cell experiments of Crandall et al. [2008], see Niessner et al. [2008]. Moreover, the overall production rate of interfacial area, E_{wn}, and especially the relative contribution of both phases to that production need to be measured. In the current model, we use an approach which allows us to provide exact values for E_{wn} along main drainage and main imbibition curves. The value of E_{wn} becomes more difficult to obtain for capillary pressure–saturation values inside the main drainage / imbibition loop. This was not needed, however, in these examples. Finally, and most difficultly, both mass diffusion length and thermal diffusion length are the most critical—as most uncertain—parameters in this numerical example. They need to be carefully measured for the interfacial-area-based model to become predictive.

In order to decide whether non-equilibrium effects with respect to mass and heat transfer are significant, Niessner and Hassanizadeh [2009c] non-dimensionalized the system of equations given by Eq. (5.86) through (5.109) and identified relevant dimensionless numbers. This is discussed in detail in Sec. 6.2.3.

5.5 Material interfaces

Natural porous media are rarely homogeneous. Pore-scale material interfaces are averaged out when going from the pore scale to the macro scale, but heterogeneities also occur on the macro scale. These macro-scale heterogeneities lead to material interfaces. These material interfaces are highly challenging for numerical modeling. While on the pore scale, there are generally smooth transitions in material properties rather than discontinuities, the situation is different on the macro-scale: properties may change abruptly from one REV to the other which causes a discontinuous distribution of material parameters. Therefore, it is essential to handle material interfaces in a physically based way by applying the valid continuity conditions.

5.5.1 Continuity conditions

For each partial differential equation describing a system, two continuity conditions have to be formulated: one to account for the continuity of fluxes and one to account for the continuity of specific state variabels (or combinations of state variables). In the following, the set of continuity conditions for an isothermal immiscible two-phase flow system as studied in Sec. 5.1 (Eq.s (5.1) through (5.7)) will be considered. The continuity conditions for the interface between two materials which are denoted by "coarse (c)" and "fine (f)" will be developed in the following.

Let us start by studying the **continuity conditions for fluxes**. For each partial differential equation, the continuity of the respective flux has to be preserved across material boundaries. For the system under study, these are the following conditions:

- continuity of the wetting-phase flux, $F_w^c = F_w^f$ at the material boundary
- continuity of the non-wetting phase flux, $F_n^c = F_n^f$ at the material boundary
- continuity of the flux of the specific fluid–fluid interfacial areas, $F_{wn}^c = F_{wn}^f$ at the material boundary.

The **continuity conditions for state variables** are harder to determine and call for a closer look at the considered system. The general quantity that needs to be continuous is specific Gibbs free energy, G_α and $G_{\alpha\beta}$. From Eq.s (3.92) and (3.93),

$$\underline{\underline{R}}^\alpha_\alpha \cdot \underline{v}_{\alpha,s} = -\phi S_\alpha \rho_\alpha \left(\nabla G_\alpha - \underline{g} \right) \tag{5.113}$$

$$\underline{\underline{R}}^{\alpha\beta}_{\alpha\beta} \cdot \underline{v}_{\alpha\beta,s} = -a_{\alpha\beta} \Gamma_{\alpha\beta} \left(\nabla G_{\alpha\beta} - \underline{g} \right), \tag{5.114}$$

it follows that the quantities for which continuity needs to be postulated highly depend on which dependencies of free energies are taken into account. In case of the equation system considered in Sec. 5.1, it can be concluded from Eq. (5.113) that these are obviously the two phase pressures, p_w and p_n. The continuity of both phase pressures implies continuity of capillary pressure. From Eq. (5.114) it follows that in case of the equation system of Sec. 5.1, specific interfacial area must be continuous as well. It is very important to note that depending on which dependencies of free energies are included in the consideration, completely different continuity requirements may result!

Summing up, continuity has to be fulfilled for the fluxes F_w, F_n, F_{wn} as well as for the state variables p_w, p_c, and a_{wn} (the continuity of p_n results from the continuity of p_w and p_c; therefore, it does not give any new information and may not be listed as an additional continuity requirement).

5.5.2 An interface condition

In this section, based on the work of van Duijn and de Neef [1998] for the classical two-phase flow model, an interface condition for material boundaries will be developed for the 2pia model introduced in Sec. 5.1. In order to maintain consistency, the interface condition for the alternative model will be derived for an S_w, p_w, p_c formulation (S_w, p_c, and p_w as primary variables) as used in that section. Of course, the interface condition can be adapted to any other choice of primary variables. The continuity condition for fluxes and p_w can be fulfilled very easily, so the following considerations will focus on the other two primary variables S_w and p_c. These two variables are connected via the $a_{wn}(S_w, p_c)$ relationship. As the interface condition can only be understood in connection with a certain numerical scheme and discretization, a vertex centered finite volume discretization as introduced in Sec. 4.3 is considered. The respective discretization at an interface between a coarse and a fine material is shown in Fig. 5.28. It is important to note that S_w, p_c, and p_w

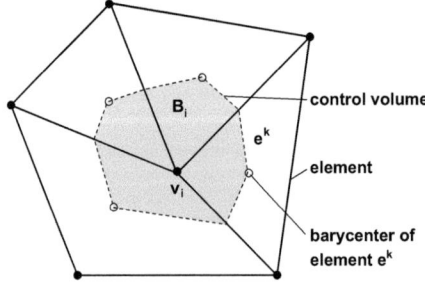

Figure 5.28: Vertex centered finite volume discretization using the 2pia model (1d consideration).

are defined on the nodes of the finite element (FE) mesh (the big dots in Fig. 5.28), while $a_{wn}(S_w, p_c)$ functions are given for the elements (the void space between the dots).

For the development of the interface condition, several simplifying assumptions are made, which are not necessary and may be relaxed in the future. Specifically, it is assumed that the following relationships among parameters for coarse material c and fine material f are valid:

- the entry pressure of the fine material is higher than that of the coarse material, $p_d^f > p_d^c$
- the intrinsic permeabilities are scalar and that of the fine material is smaller, $K^f < K^c$
- equal porosities $\phi^f = \phi^c$
- the bounding capillary pressure–saturation curves in both materials can be parametrized using the Brooks and Corey model with the same λ-parameter, $\lambda^f = \lambda^c$.

These conditions also give a more precise definition of the terms "fine" and "coarse". From the above condition for entry pressures, it follows that $a_{wn}^f(S_w) > a_{wn}^c(S_w)$ because of increased curvature in small pores for the same saturation (see also Grant and Gerhard [2007]). The inequality $K^f < K^c$ acts in the same direction.

Fig. 5.29 shows the $p_c(S_w)$ functions of two points which are located directly at the interface, one on the coarse side and the other one on the fine side as well as the functions $a_{wn}(S_w)$ for a fixed value of p_c (right hand side). The capillary pressure–

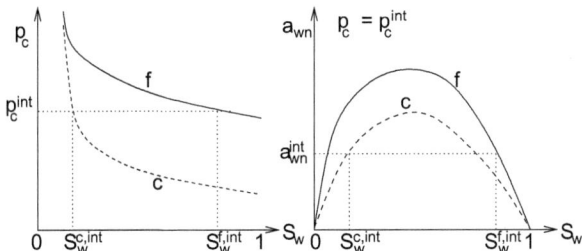

Figure 5.29: Graphs of $p_c(S_w)$ (left) for two points located directly at a material interface and $a_{wn}(S_w)$ (right) for a fixed capillary pressure for fine and coarse material.

saturation functions on the left hand side of Fig. 5.29 indicate the pathlines in the capillary pressure–saturation plain for two points. They were obtained by projecting the three-dimensional pathlines from the $a_{wn}(S_w, p_c)$ surface onto the capillary pressure–saturation plane. The two points are chosen such that they are located on the interface, but one belonging to the fine material and the other one to the coarse material. It can be clearly seen that a continuity in capillary pressure across the

181

material interface implies a discontinuity in saturation across that interface. The saturation jump is given by the difference $S_w^{f,int} - S_w^{c,int}$. On the right hand side of Fig. 5.29, $a_{wn}(S_w)$ functions are shown for coarse and fine material for a fixed value of capillary pressure that corresponds to the capillary pressure p_c^{int} directly at the interface. It is obvious that the fulfillment of the continuity condition for p_c and a_{wn} provides two possible saturation values within each material. In order to decide which saturation is correct we can make use of the condition that saturation is always higher in finer than in coarser materials, $S_w^f > S_w^c$, which can be seen by comparing the capillary pressure–saturation functions for the coarse and fine material in Fig. 5.29.

The above considerations make it possible to actually formulate the interface conditions for material discontinuities. The procedure is closely related to the works of van Duijn and de Neef [1998] and Bastian [1999]. The first step (a preconditioning step) is to calculate the minimum specific interfacial area for each node. This is done by evaluating the specific interfacial area–capillary pressure–saturation functions for each node i with respect to all elements k (their barycenter is $\underline{x}^k$) which have i as a corner (this set of elements is denoted by $E(i)$),

$$a_{wn,i}^{min} = \min_{k \in E(i)} a_{wn}\left(\underline{x}^k, S_{w,i}, p_{c,i}\right). \tag{5.115}$$

In order to model the saturation discontinuity, for each node i, a so-called virtual saturation is calculated which is based on the evaluation of the specific interfacial area–capillary pressure–saturation relationships for all elements $k \in E(i)$ which have i as a corner. If the calculated specific interfacial area is equal to the minimum interfacial area, the virtual saturation is the same as the saturation of node i. This means that node i is either located in the interior of a material or at the interface and the considered element k is an element of the coarsest considered material located at that interface. If, for a primary drainage process, the entry pressure p_d of element k is not reached, the virtual wetting-phase saturation is 1. In all other cases, the virtual saturation is calculated such that it lies on the $a_{wn}(S_w, p_c)$ surface of element k. This procedure can be expressed by the following formula:

$$\tilde{S}_{w,i,k} = \begin{cases} S_{w,i} & \text{if } a_{wn}\left(\underline{x}^k, S_{w,i}, p_{c,i}\right) = a_{wn,i}^{min} \\ 0 & \text{if } p_{c,i} < p_d\left(\underline{x}^k\right) \\ \tilde{S}_{w,i} & \text{else, where } \tilde{S}_{w,i} \text{ solves } a_{wn}\left(\underline{x}^k, \tilde{S}_{w,i}, p_{c,i}\right) = a_{wn,i}^{min} \end{cases} \tag{5.116}$$

If the considered process is not a primary drainage process, but if instead, the main loop bounded by main imbibition and main drainage curve is reached, the second condition must not be evaluated.

As indicated in the right-hand picture of Fig. 5.29, there are still two possible saturation values for the virtual saturation. These are the two intersections of the curve "c" with the horizontal dotted line $a_{wn} = a_{wn}^{int}$. Therefore, when solving the equation system, the coarse saturation is chosen as primary variable and this not only within the coarse material, but in the total domain. Then, the reconstructed virtual saturation must be the fine saturation. The correct fine saturation out of the two possible values, however, can be determined from the coarse saturation as $a_{wn}^f(S_w) > a_{wn}^c(S_w)$ has to be valid for a given saturation and $S_w^f > S_w^c$ has to be fulfilled. These conditions together with the interface condition uniquely define the value of the saturation in the fine material.

5.6 Gas–water processes in a fracture–matrix system

Gas–water processes in fracture–matrix systems occur in a number of geological, biological, and technical systems. Examples for such systems are:

- **Geological systems**
  A typical example of gas–water flow processes occuring in a geological fractured porous medium is the flow of air and water in the unsaturated zone of the subsurface in case of fractured rock. These flow processes are extremely important for drinking water supply and agricultural water management. Another crucial example are flow and transport processes occuring in deep atomic or chemical waste storage sites. In Sweden, the most favored solution is deep underground storage in granite which generally represents a fractured porous medium. Atomic and chemical waste potentially produce gas (e.g. hydrogen) which may form a separate phase if pressure drops or temperature increases and subsequently lead to gas–water processes in depths of several hundreds of metres (Reichenberger et al. [2004]), where otherwise single (water-) phase flow conditions prevail. A gas phase may then travel relatively quickly up to the surface due to buoancy forces and preferential flow paths provided by the fracture system.

- **Technical systems**
  A technical system of increasing importance is the PEM fuel cell as it may serve as environmentally friendly mobile energy supply. Within such a PEM fuel cell, a controlled hydrogen-oxygen reaction takes place in an aqueous environment. Thus, these processes belong to the class of gas–water processes. With respect to the PEM fuel cell itself, several parts with different functionality can be distinguished: examples are the gas distributor where free flow of gas takes place and the diffusion layer where porous media flow occurs (Ochs et al. [2007]). The gas distributor can be interpreted as a fracture that is connected to a porous medium (the diffusion layer).

- **Biological systems**
  The exchange and interaction between blood vessels and interstitial space, e.g. in the human body, is a vital process and extremely important for the delivery of therapeutic agents to the places where they are meant to operate. One application of these agents is the treatment of tumors which are located within

the interstitial space and which need to be reached by therapeutic agents via the blood vessels (Smith and Humphrey [2006]). Generally, the blood vessels may be interpreted as fractures in the surrounding "porous medium", the interstitial space. In contrast to the other applications, this biological fracture–matrix system is a single-phase system.

Flow velocities in fractures are generally orders of magnitude larger than in the matrix. This leads to the fact that very often, non-equilibrium situations are encountered within fractures even if within the porous matrix, chemical equilibrium may be a reasonable assumption. Also, a fracture–matrix system represents a heterogeneous domain. In such a system, drainage and imbibition processes may take place locally depending on the heterogeneous structure of the domain. This means e.g. that even if the whole domain undergoes drainage, local imbibition may occur. Using the classical two-phase flow approach, hysteretic capillary pressure–saturation functions have to be provided. Thus, the alternative approach is useful for the application to fracture–matrix systems mainly in two ways:

1. Using a two-phase two-component model, non-equilibrium situations occurring in the fractures can be accounted for in a natural way as specific interfacial area is a known parameter.

2. Local hysteresis can be captured without providing scanning curves, but by allowing movement on the specific interfacial area–capillary pressure–saturation surface.

Based on the work of Niessner and Hassanizadeh [2008, 2009a,b] for porous media, these two steps have been performed in the frame of two Dipl.-Ing. theses: Nuske [2009] extended a micro-scale model of Jakobs [2004] and Reichenberger et al. [2004] for a single fracture to perform micro-scale simulations that yield specific interfacial area, capillary pressure, and saturation. Based on Nuske's data, Faigle [2009] modeled gas–water processes in a macro-scale fracture–matrix system. The following discussions and results in principal follow the line of these two Dipl.-Ing. theses.

Conceptually, fractures are often treated as porous media systems with high porosity and permeability as well as low entry pressure. In that case, it is necessary to provide porosity, intrinsic permeability, as well as the specific interfacial area–capillary pressure–saturation surface for the fractures. Therefore, the following discussion comprises two steps, see also Fig. 5.30:

1. Calculation of parameters and constitutive relationships for fractures by use of a micro-scale model for single fractures (Nuske [2009]), see Sec. 5.6.1.

2. Macro-scale simulation based on a two-phase two-component model including specific interfacial area to account for hysteresis and kinetic interphase mass transfer (Faigle [2009]), see Sec. 5.6.2.

5.6.1 Calculation of effective parameters

First of all, fractures need zo be geometrically represented. Therefore, the irregular fracture walls need to be discretized, see picture a) of Fig. 5.31. In the following, a raster element description of the fracture walls will be considered as shown in picture b). Once a discretization is found, effective parameters (porosity, intrinsic permeability, and a constitutive relationship between specific interfacial area, capillary pressure, and saturation) can be calculated. The exact formulas are provided in the following.

5.6.1.1 Porosity

Porosity of fractures can either be chosen to be equal to unity arguing that there is actually no porous medium within the fracture or by following an analogous procedure as for porous media porosity and choosing its value such that it represents the volume fraction of void space within the fracture.

5.6.1.2 Intrinsic permeability

As a raster element discretization of the fracture walls was chosen a parallel plate model can be applied within each raster element. The strategy for the calculation of intrinsic permeability is then to calculate its value for each raster element and subsequently average it over the whole fracture.

The starting point for this calculation is the x-component of the three-dimensional Navier–Stokes equation:

$$\varrho g_x - \frac{\partial p}{\partial x} + \mu \left(\frac{\partial^2 v_x}{\partial x^2} + \frac{\partial^2 v_y}{\partial y^2} + \frac{\partial^2 v_z}{\partial z^2} \right) = \varrho \frac{dv_x}{dt} \quad (5.117)$$

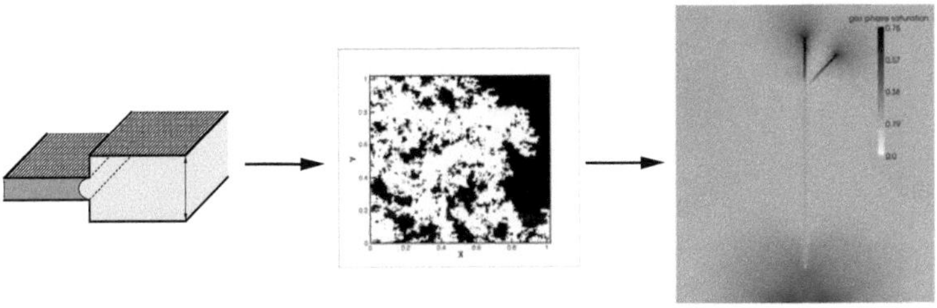

Figure 5.30: From the conceptual model of a single fracture (left hand side; here, a cutout is shown) over micro-scale simulation to macro-scale simulation.

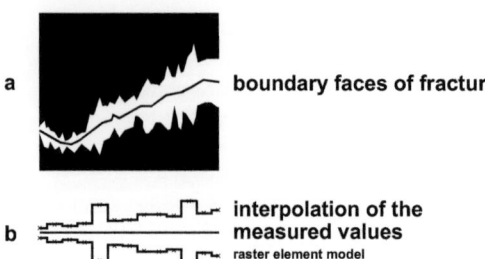

Figure 5.31: Different representations of the fracture surfaces (modified from Jakobs [2004]).

For incompressible, isothermal, laminar, horizontal, stationary, creeping flow in a raster element with hydraulically smooth fracture walls, this equation simplifies to

$$v_x = \frac{a^2}{8\mu}\frac{\partial p}{\partial x}\left(\frac{4}{a^2}z^2 - 1\right), \tag{5.118}$$

where z is a direction perpendicular to the fracture wall, $z = 0$ in the middle of the fracture, a is the distance between the two opposing walls of the fracture, the so-called fracture *aperture*, and μ is the viscosity of the fluid.

Using dimensional analysis, Jakobs [2004] found this relation to be true under two constraints:

1. The raster element length is much bigger than the mean fracture aperture: $\Delta x \gg \bar{a}$.

2. The raster element length is smaller than the correlation length of the fracture. This constraint is required in order to resolve the geometry.

The mean flow velocity in one raster element can be obtained by integrating over the fracture aperture and dividing by the aperture. Flow through a raster element of base $w \cdot w$ is thus

$$\dot{V} = A\bar{\underline{v}} = w\,a\,\bar{\underline{v}} \tag{5.119a}$$

$$\dot{V}_x = -w\frac{a^3}{12\mu}\frac{\partial p}{\partial x}, \tag{5.119b}$$

with $A = w \cdot a$ denoting the cross section of one raster element. Due to the aperture a being to the third power, Eq. (5.119b) is called cubic law. Comparison with Darcy's law

$$\dot{V} = -wa\underline{\underline{K}}_f \cdot \nabla h = -wa\frac{\varrho g}{\mu}\underline{\underline{K}} \cdot \nabla h \tag{5.120}$$

reduced to one spatial dimension

$$\dot{V}_x = -wa\frac{\varrho g}{\mu}K\frac{\partial}{\partial x}\frac{p}{\varrho g} = -wa\frac{1}{\mu}K\frac{\partial p}{\partial x} \qquad (5.121)$$

suggests that the equivalent intrinsic permeability of one raster element in a fracture with the aperture a be

$$K = \frac{a^2}{12}, \qquad (5.122)$$

assuming that Darcy's law is valid for flow in fractures and the above mentioned constraints regarding the cubic law are met.

5.6.1.3 Specific interfacial area–capillary pressure–saturation surface

In order to determine the specific interfacial area–capillary pressure–saturation surface it needs to be discussed first how, from a given occupancy of the fracture aperture field, the parameters saturation, capillary pressure, and specific interfacial area can be determined. In a second step, it will be discussed how the occupancy of the fracture apertures can actually be determined using an invasion model.

Calculation of saturation Calculation of saturation of a phase α is trivial as it can simply be calculated using its definition, as the volume fraction of the fracture that is occupied by phase α. However, two assumptions regarding the saturation are made: (1) there is a residual water saturation S_{wr} of 20% because water may be trapped within the rough fracture walls and (2) there is a thin water film on the fracture wall.

Calculation of capillary pressure According to the Young-Laplace equation given in Eq. (2.16), micro-scale capillary pressure is proportional to surface tension and inversely proportional to the main radius of curvature. A contact angle of $0°$ is assumed which is in line with the observed water film on the fracture walls. This in connection with the requirement discussed above, $\Delta x \gg \bar{a}$, results in a cylindrical shape of the interface (see Fig. 5.32),

$$p_c = \sigma_{wn}\left(\underbrace{\frac{1}{r_{xy}}}_{0} + \underbrace{\frac{1}{r_z}}_{2/a}\right) \qquad (5.123\text{a})$$

$$p_c = \sigma_{wn}\frac{2}{a}, \qquad (5.123\text{b})$$

where r_{xy} denotes the radius of curvature in x and y direction, and r_z is the radius of curvature in z-direction with $r_z = a/2$. For a cylinder, $r_{xy} = \infty$. Therefore, its

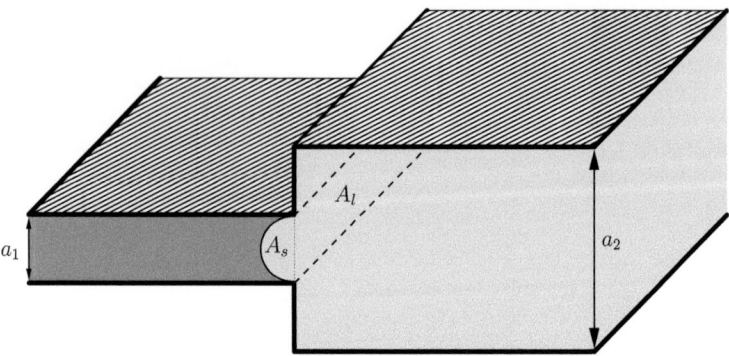

Figure 5.32: Simplified interface geometry. The left raster element is occupied by wetting phase w and the right raster element is occupied by the non-wetting phase n.

reciprocal occuring in the first term in brackets is zero. From the formula obtained in Eq. (5.123b), the smallest aperture that is occupied by gas for a given set of boundary conditions p_n and p_w can be determined,

$$a_{min} = 2\frac{\sigma_{wn}}{p_n - p_w}, \tag{5.124}$$

where $p_c = p_n - p_w$ has been taken into account. This means, for a given set of boundary conditions p_n and p_w, all apertures $a \geq a_{min}$ can be occupied by gas.

Calculation of specific interfacial area In the following, only the interfacial area due to the meniscus between gas and water phase which is located between two raster elements as shown in Fig. 5.32 is taken into account. This is only a first order approximation, as there are obviously also interfaces located in the corners of the fracture elements.

As explained above, a cylinder is a good approximation of the shape of the meniscus between two adjacent raster elements. Using this approximation, only the lateral surfaces of the cylinder are "counted" as interfaces (A_l in Fig. 5.32). This approximation is justified if the ratio of A_s/A_l is small (see Fig. 5.32).

Considering these issues, the total interfacial area within a fracture is calculated as follows:

$$A_{wn} = \sum_i 0.5 \, \Delta x \, a_i \, f_{ij} \quad \text{where} \quad \begin{cases} f_{ij} = 0 & \text{if } i, j \text{ occupied by same phase} \\ f_{ij} = 0.5 \, \pi & \text{if } i, j \text{ occupied by different phases} \end{cases} \tag{5.125}$$

with A_{wn} as the interfacial area between wetting and non-wetting phase. Throughout this section, equidistant raster elements of edge length Δx are chosen. In order to obtain the total interfacial area within a fracture, the areas between different raster elements are summed over the raster elements i. Therefore, each interface is counted twice which is compensated by multiplication with a factor of 0.5.

The described algorithm determines interfacial area, but the quantity needed in the alternative approach is specific interfacial area. Therefore, the calculated total interfacial area A_{wn} is divided by the total volume V_{fract} of the fracture which is determined as the sum of the raster element volumes over all raster elements i,

$$V_{fract} = \sum_i \Delta x \, \Delta x \, a_i . \tag{5.126}$$

Invasion model—determination of occupancies In the invasion model, a setup is considered where a fracture is initially fully saturated with water. Gas invades the fracture from one boundary, water may invade subsequently from the opposite boundary. The difference of phase pressures at two opposite boundaries is identified as capillary pressure as is done in (static) experiments to determine capillary pressure–saturation relationships. This capillary pressure value determines which raster elements can be occupied by the invading gas phase.

In order to obtain lots of different points on the specific interfacial area–capillary pressure–saturation surface, several different drainage and imbibition processes need to be modeled. Therefore, a switch from gas phase invasion (drainage) to water phase invasion (imbibition) and vice versa is done.

For invasion of a raster element, three criteria have to be met

1. The considered raster element has to be available according to the Young-Laplace equation (Eq. (2.16)).
2. There has to be supply of invading phase.
3. There has to be a way out for the displaced fluid.

If one raster element of the fracture can be invaded it will be completely invaded. In the case of drainage, this complete invasion corresponds to a gas saturation of 0.8 within the raster element (due to the presence of water in corners of the rough surface of the fracture walls). In the case of imbibition, a water saturation of one is reached within a raster element after invasion.

The first criterion yields a criterion for the aperture,

$$a_{cut} \gtrless 2\frac{\sigma}{p_c} , \tag{5.127}$$

where the greater-than-sign applies to drainage and the smaller-than-sign to imbibition, respectively and a_{cut} means cut-off aperture. The cut-off aperture separates those raster elements that can be invaded by gas from those which cannot be invaded. In the case of imbibition, all raster elements that are *smaller* than a_{cut} are available for water, whereas in the case of drainage, all raster elements *larger* than a_{cut} are available for gas.

Not all raster elements that are available—according to Eq. (5.127)—are also accessible. If the availability was the only criterion, the relationship between capillary pressure and saturation was unique.

The second criterion is met if at least one of the four neighboring raster elements of the considered raster element is filled with invading phase *and* is connected to its reservoir. This situation is illustrated in Fig. 5.33 where water is represented by dark grey and gas by light gray: raster elements $\textcircled{n}_i$ are the neighbors of the considered raster element $\textcircled{c}$. Raster element $\textcircled{n}_4$ is connected to the gas reservoir, as indicated by the solid line. Connectivity is determined by following connected elements from the considered reservoir: in a first step, the connectivity of the invasion boundary is set to one. Subsequently, for each neighboring raster element, it is tested whether the element is filled with invading phase. If it is, its connectivity to the invasion reservoir is also set to one. This routine is continued for all raster elements until there is no change in the number of connected raster elements any more.

The same routine is applied for the displaced phase with the reservoir opposing the invasion reservoir and using the other saturation. The only difference is that the connectivity has to be re-calculated after a raster element has been occupied.

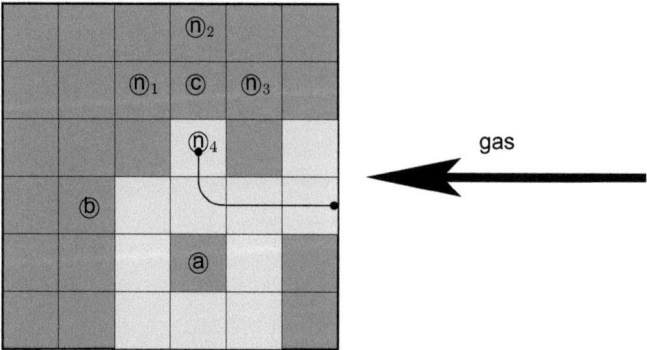

Figure 5.33: Illustration of a drainage process. Water phase is dark gray and gas phase is light gray.

The assumption of a water film on all fracture walls leads to the fact, that the water phase is always connected throughout the domain. This is consistent with assuming a strongly water-wet system with a contact angle of 0°. Tokunaga and Wan [1997]

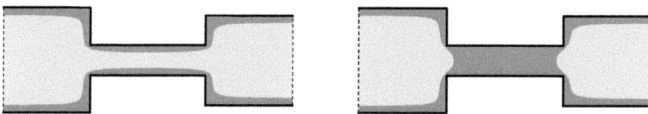

Figure 5.34: Water (dark) and gas (light) in three fracture elements. Left hand side: the raster element in the middle cannot be invaded by wetting phase; right hand side: invasion by non-wetting phase is possible (Tokunaga and Wan [1997]).

tried to quantify the flow occuring in these films experimentally and came to the conclusion that *film flow* on fracture walls is an important flow mechanism during drainage. If film flow also took place during imbibition, this would imply that a raster element could be invaded through a film. For this to take place, the film would have to "detach" from the wall and thus overcome adhesive forces, which are known to be strong. In Fig. 5.34, a sketch of these processes is shown. No matter what the capillary pressure, in the left configuration, the raster element in the middle cannot be invaded by the water phase. In the right configuration, contrarily, the raster element in the middle can be invaded by gas phase if capillary pressure is sufficiently high.

In this section, fracture elements are either filled by water ($S_w = 1$) or gas ($S_w = 0.2$) and film flow is not considered explicitly. Instead, the above mentioned processes are incorporated by allowing water phase to leave a raster element, as it is always connected to other raster elements through films. Therefore, a continuous gas phase can force water through this film even if there is no connection of fully water-filled fracture elements to the water reservoir. Thus, the third criterion is always met in the case of drainage (*no-trapping assumption for water* [Joekar-Niasar et al., 2008]).

Joekar-Niasar et al. [2008] explain that the decision whether the wetting phase stays mobile or not can be made based on experimental data: if a phase stays mobile, a specific interfacial area versus saturation plot would show a maximum in between the boundary of the data range. If trapping occured, contrarily, specific interfacial area would increase monotonically with decreasing wetting-phase saturation. Later, Joekar-Niasar et al. [2009] developed a pore-network model based on a micro-model experiment. As the experimental model shows a maximum in $a_{wn}(S_w)$, they deduced that the wetting phase is free to leave any cell. By using this assumption in the numerical model, very good agreement between micro-model observation and pore-network modeling was achieved.

If the drainage process is not primary drainage not all raster elements occupied by gas are necessarily connected to their reservoir. A schematic imbibition process is shown in Fig. 5.35. Water invades over the left boundary. Again, invading means that connectivity is determined starting from this boundary. As gas is only mobile when connected to its reservoir (right boundary) the patches marked with ⓓ and ⓔ can never be invaded by water in this configuration, independently of capillary pressure: there, the gas is trapped. In pore-network models, flow through porous

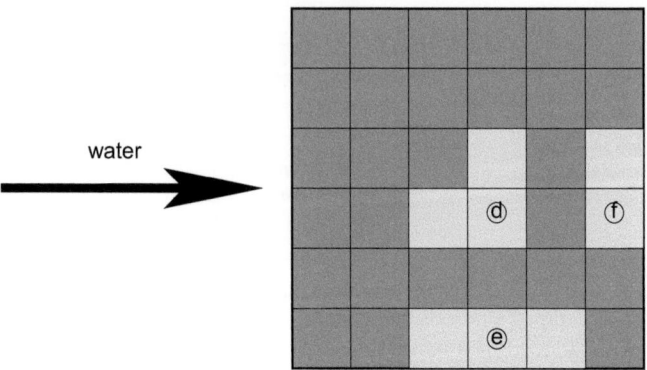

Figure 5.35: Schematic of the connectivity situation for an imbibition process. Water phase is dark gray and gas phase is light gray.

media is conceptualized as flow through a network of *pore bodies* connected by *pore throats*. Applying this terminology, the described model can be interpreted as a pore network model consisting solely of pore bodies.

A model run of the invasion model starts with all fracture elements being fully saturated with water. Gas phase invades over one boundary by means of increasing capillary pressure (the difference between phase pressures at the left and right boundary). New fracture elements will be occupied by gas phase—according to the above described invasion criteria—until no change in the number of gas occupied fracture elements occurs any more. Then, the equilibrium state is reached and the properties of interest (S_w, a_{wn}) are calculated. Subsequently, the capillary pressure is further increased and new equilibrium states are calculated. The simulation continues on the primary drainage curve until the resulting change in water saturation for a given change in capillary pressure is smaller than a predefined limit. If this limit is reached the model is switched to imbibition. During imbibition, capillary pressure is lowered stepwise. For each pressure step, the invasion rules are applied as described above. Thus, primary drainage and main imbibition curves can be obtained. Increasing capillary pressure again until reaching the previously determined maximum capillary pressure results in the main drainage curve. These two curves make up the bounding curves of the main hysteresis loop.

Starting from the main drainage curve, scanning curves can be obtained by changing the invasion mechanism from drainage to imbibition. Each scanning drainage curve is continued until the bounding curve is reached and the invasion mechanism is changed back to drainage. By this procedure, p_c, S_w, a_{wn} data is obtained for each equilibrium state in the main hysteresis loop. Scanning curves can also be started from the primary drainage curve. The lower boundary is in this case a capillary pressure of zero.

5.6.2 Macro-scale simulation of two-phase flow in a fracture–matrix system

For the macro-scale simulations, two setups as shown in Fig. 5.36 are studied: fully water-saturated gas is infiltrated into a two-dimensional, vertical, and initially highly water-saturated domain of size 8 m times 10 m. One setup consists of a homogeneous porous medium while the other one contains two intersection fractures. Neumann no flow boundary conditions are specified on both lateral sides and at the bottom, with the exception of an infiltration zone between 3 m to 5 m from the lower left corner. At the top of the model domain, a Dirichlet-boundary is chosen, values of the unknowns will be given later in this section. The coefficients of the

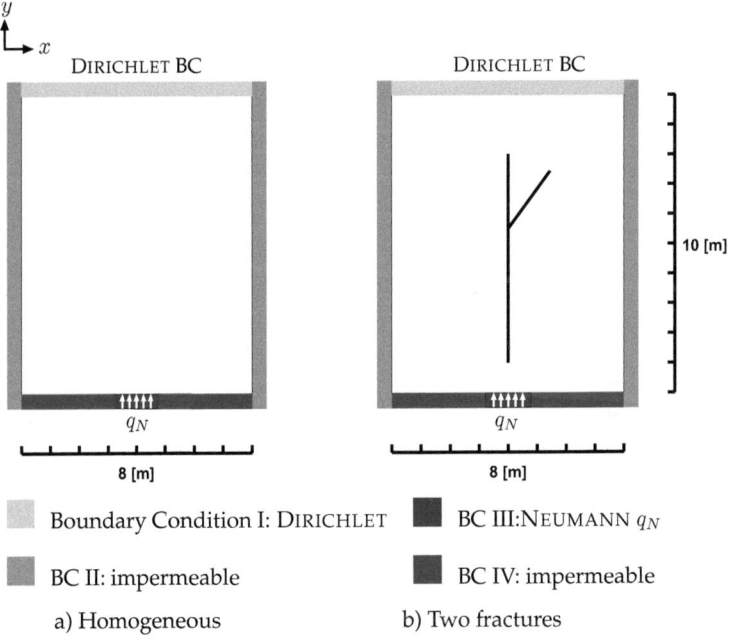

Figure 5.36: Sketches of the applied model setups (left hand side: homogeneous porous medium, right hand side: porous medium with two fractures).

specific interfacial area–capillary pressure–saturation surface for the porous matrix are taken as obtained by Joekar-Niasar et al. [2008] using a pore-network model and those of the fracture are taken as obtained by Nuske [2009] using an invasion model, see Tab. 5.4. Bounding drainage and imbibition curves for both fracture and matrix

	$a_{00}\,[\tfrac{1}{m}]$	$a_{01}\,[\tfrac{1}{m}]$	$a_{10}\,[\tfrac{1}{m\cdot Pa}]$	$a_{11}\,[\tfrac{1}{m\cdot Pa}]$	$a_{20}\,[\tfrac{1}{m}]$	$a_{02}\,[\tfrac{1}{m\cdot Pa^2}]$
Matrix	688.886	4167.65	-0.185	-0.039	-4058.7	$1.058 \cdot 10^{-5}$
Fracture	-73.35	391.95	0.0058	-0.0232	-301.3818	$-2.569 \cdot 10^{-8}$

Table 5.4: Coefficients of the specific interfacial area–capillary pressure–saturation surface for matrix (Joekar-Niasar et al. [2008]) and fracture (Nuske [2009]).

are parametrized by van Genuchten functions.

For this step, three different models will be compared in the following:

1. A classical two-phase two-component model denoted by *2p2c* (see Sec. 2.3.1.2). This model can only reproduce mass fractions equal to the equilibrium mass fractions. The capillary pressure–saturation relationship is parametrized using a van Genuchten function for the primary drainage curve. Thus, this model is not able to describe capillary hysteresis.

2. An extended two-phase two-component model (*2p2cia4*) which includes non-equilibrium kinetic mass transfer over the phase-interfaces but is only valid for primary drainage. Thus, a specific $p_c(S_w)$ curve is imposed. In that case, no interfacial area balance equation needs to be included and a direct relationship between specific interfacial area and saturation can be given. The "4" indicates that only four partial differential equations need to be solved.

3. The interfacial-area-based approach as discussed in Chap. 3 (*2p2cia5*). This model can account for chemical non-equilibrium and capillary hysteresis. Here, all five (5) partial differential equations are solved.

The set of initial values (Tab. 5.5) has to be chosen according to the set of primary variables. Boundary conditions are given according to Fig. 5.36 where Dirichlet is chosen along the top side, and all other boundaries are of Neumann type with

$$q_{N,n}(x,y) = \begin{cases} -0.005\ \tfrac{kg}{m^2 s} & \text{if on bottom boundary and}\quad 3\,m \leq x \leq 5\,m \\ 0 & \text{otherwise.} \end{cases} \quad (5.128)$$

Initial values (I.C.) and values of the Dirichlet boundary condition (B.C.) are the same and given in Tab. 5.5.

Verification study for the conceptual models A homogeneous setup allows general validation of the three different model concepts. Therefore, the state of the system after 100 seconds is compared, see Fig. 5.37. The encountered differences are

	I.C. and B.C.	model
Wetting-phase pressure p_w [Pa]	$10200 + 10 \cdot \rho_w g$	2p2c, 2p2cia4, 2p2cia5
Non-wetting phase saturation S_n [-]	0.05	2p2c, 2p2cia4, 2p2cia5
Mass fraction X_w^a [-]	$\frac{p_n}{H_w^a}$	2p2c, 2p2cia4, 2p2cia5
Mass fraction X_n^w [-]	$\frac{p_{w,sat}}{p_n}$	2p2cia4, 2p2cia5
Capillary pressure p_c [Pa]	$p_c^{dr}(S_w)$	2p2cia5

Table 5.5: Initial values and Dirichlet boundary conditions for the three compared models.

predominantly caused by the different relative permeability functions of the respective models (quadratic versus van Genuchten function). For comparison, therefore, the relative permeability function of the alternative approach, $k_{r\alpha} = S_\alpha^2$, is used in all models which yields good agreement (Fig. 5.37(d)).

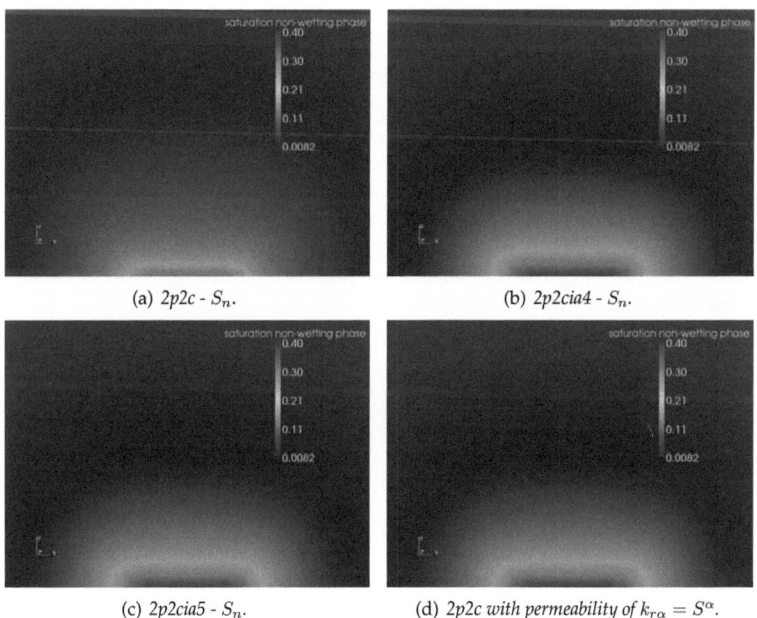

(a) 2p2c - S_n.

(b) 2p2cia4 - S_n.

(c) 2p2cia5 - S_n.

(d) 2p2c with permeability of $k_{r\alpha} = S^\alpha$.

Figure 5.37: Comparison of the different model concepts for the homogeneous case.

Fig. 5.38(a) indicates a short imbibition process at the beginning of the simulation, thus capillary pressure does not follow the primary drainage curve. The reason is that when entering the domain, the propagating gas front compresses the water

ahead, thus creating a small imbibition zone ahead of the front.
From Fig. 5.38, it becomes clear that the assumption of chemical equilibrium seems

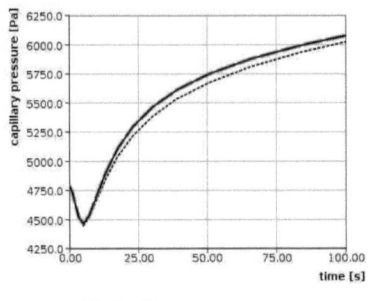
(a) Capillary pressure p_c.

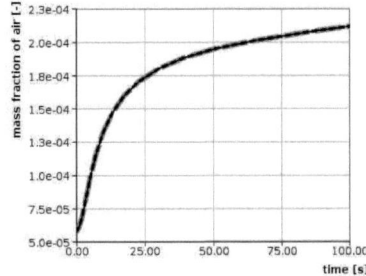
(b) Mass fraction of air in the water phase X_w^a.

Figure 5.38: Capillary pressure and mass fraction X_w^a as a function of time at a node.

appropriate for the slow homogeneous case. This means that the mass fractions obtained by the extended models (2p2cia4, 2p2cia5) are very close to the equilibrium mass fractions given by the standard approach, Fig. 5.38(b). Although mass transfer due to non-equilibrium is recognizeable in Fig. 5.39(a), the deviation from the equilibrium mass fraction is marginal: the mass transfer term includes a diffusion coefficient which is five orders of magnitude larger than the one for the component air. Significant effects of non-equilibrium are only detectable for the component air, see Fig. 5.39(b). Areas below equilibrium, indicated by a positive sign of the mass transfer, are only observable in the vicinity of the infiltration zone.

Fig. 5.40 shows the saturation of the non-wetting phase (grey) as well as the non-

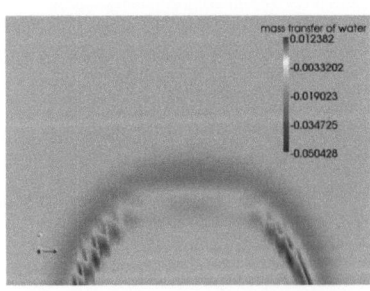
(a) Mass transfer of water into gaseous phase $I_n^W [\frac{kg}{sm^3}]$.

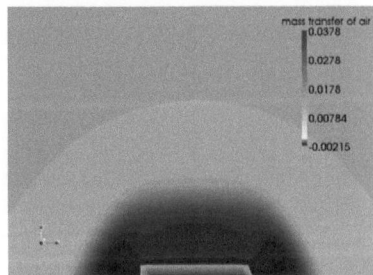
(b) Mass transfer of air into water phase $I_w^A [\frac{kg}{sm^3}]$.

Figure 5.39: Mass transfer after 50 seconds.

Table 5.6: Computational expenses for a 100 second simulation run.

model concept	timesteps	computational time
2p2c	14	80 s
2p2cia4	14	158 s
2p2cia5	15	228 s

equilibrium mass transfer of air into water I_w^A (black) along a one-meter-cutout perpendicular to the infiltration zone. This cutout is marked by the white line on the right picture of Fig. 5.40. An infiltration rate of $q_N = -0.005 \frac{kg}{m^2 s}$ is represented by a solid line while the dotted line marks $q_N = -0.007 \frac{kg}{m^2 s}$ and the dashed line $q_N = -0.01 \frac{kg}{m^2 s}$. With increasing infiltration rate the gas phase velocity also increases. In that case, non-equilibrium effects are more pronounced.

Tab. 5.6 lists the time to complete the simulation of 100 seconds as well as the nec-

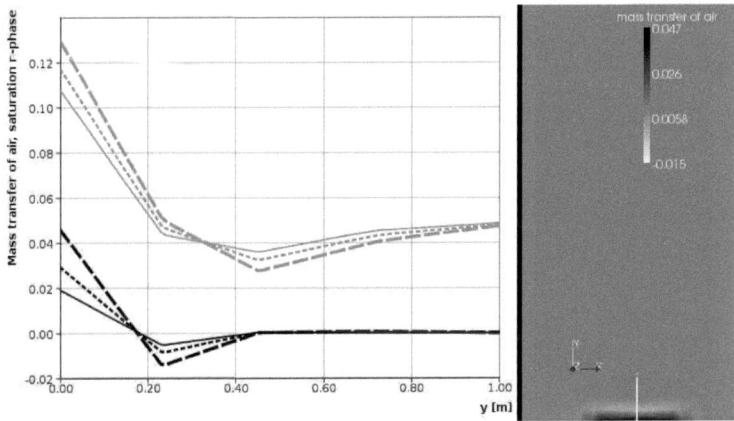

Figure 5.40: Mass transfer depending on infiltration rates (black in $\left[\frac{kg}{m^3 s}\right]$) and saturation of the gaseous phase (grey, in [-]).

essary timesteps. As expected, the solution of more partial differential equations increases the computational demand. Both 2p2cia4 and 2p2cia5 models require much more NEWTON-iterations to reach convergence which explains increased computational demand without an increase in timesteps.

To study the influence of buoyancy on the different processes, a fracture–matrix system as shown in Fig. 5.41 is chosen. This setup poses high demands on gridding. In Fig. 5.42, mass transfer occuring at two nodes with comparable pressure gradients

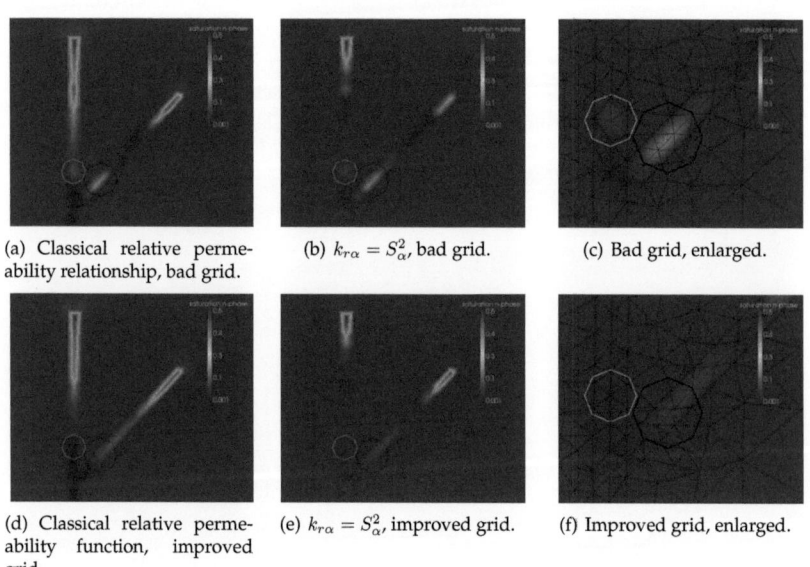

Figure 5.41: Grid dependence of the solution. A simulated time of 6 seconds is compared.

(a) Classical relative permeability relationship, bad grid.
(b) $k_{r\alpha} = S_\alpha^2$, bad grid.
(c) Bad grid, enlarged.
(d) Classical relative permeability function, improved grid.
(e) $k_{r\alpha} = S_\alpha^2$, improved grid.
(f) Improved grid, enlarged.

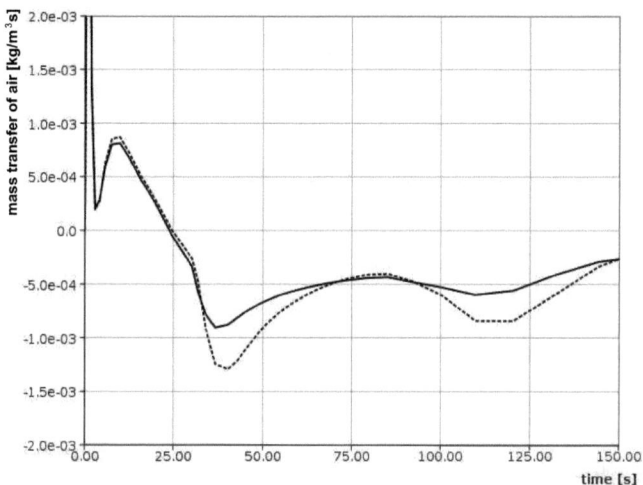

Figure 5.42: Mass transfer of air over simulation time for a node in the vertical (dotted line) and inclined (solid line) fracture.

in both parts of the fracture system is shown. As may be expected, non-equilibrium mass transfer is detected to be higher in the vertical than in the inclined fracture due to the higher velocity. Note that the results are highly dependent on the grid.

In this section, a methodology was presented that allows to model kinetic interphase mass transfer and hysteresis in two-phase flow in fracture–matrix systems. Constitutive relationships are obtained from micro-scale simulations and, based on these functions, of a single fracture and macro-scale simulations for fracture–matrix systems are performed.

5.7 Determination of parameters

In this section, the parameters of the macro-scale model based on the interfacial-area-based approach will be discussed in a more practical sense. It will be shown how the relevant model parameters and constitutive relationships can be obtained from experiments, pore-network models, or Lattice-Boltzmann simulations.

5.7.1 Overview of the methods

Parameters for the interfacial-area-based model, and specifically interfacial areas themselves, can be obtained **experimentally** using the following techniques:

a. Interfacial tracers.
   A special kind of interfacial tracer (surfactant), which is not soluble in the non-wetting phase, is dissolved in the water phase. Such interfacial tracers form mono-molecular layers on the fluid–fluid interface and the remaining amount of tracer stays in solution in the wetting phase. Now, if the interfacial area increases, tracer molecules will move from the bulk water phase to the interface and the concentration of tracer in the water phase decreases. A decrease of interfacial area, contrarily, will lead to an increase in tracer concentration in the water phase.

   In order to determine the $a_{wn}(S_w, p_c)$ relation, non-wetting phase pressure is changed stepwise which leads to capillary pressure changes. This leads to stepwise changes in saturation and in specific interfacial area. As a result, tracer molecules move from the bulk wetting phase to the wn-interface (or vice versa) and the concentration of the surfactant in the wetting phase changes. Measuring the concentration of the surfactant in the wetting phase thus allows to estimate interfacial area, see e.g. Chen and Kibbey [2006]. Note that other kinds of interfacial tracers exist (e.g. surfactants that are only soluble in the non-wetting phase) and can also be used for experimental determination of interfacial areas.

b. Imaging techniques.
   By digital imaging, the geometric phase distribution and the location of the interfaces in a transparent microfluidic cell can be recorded. Subsequent image analysis gives values for both saturation and interfacial area, see e.g. Pyrak-Nolte et al. [2008], Chen et al. [2007]. While this procedure is relatively easy, it only allows for a two-dimensional analysis of the system.

c. Micro tomography.
   In micro tomography, small samples of a three-dimensional porous medium are analyzed in a synchroton. A full three-dimensional image of the porous medium and fluid distribution can be reconstructed. Through image analysis, pore occupancies can be identified and saturation, capillary pressure, and specific interfacial area can be calculated, see e.g. Wildenschild et al. [2002].

Concerning **numerical investigations** of Hassanizadeh and Gray's conjecture, there are also two main directions of numerical techniques: pore-network models and Lattice-Boltzmann methods.

a. Pore-network models.
   Static pore-network models have served as an alternative means of constructing p_c–S_w–a_{wn} relationships. Reeves and Celia [1996] calculated specific inter-

facial area–capillary pressure–saturation surfaces, both for drainage and imbibition. While their model gave a difference between these two surfaces, Held and Celia [2001] showed that including effects of snap-off and local fluid configurations during imbibition, the difference between the two surfaces can be reduced down to a very small value. Finally, Joekar-Niasar et al. [2008] studied different trapping assumptions and obtained both interfacial area–capillary pressure–saturation surfaces and interfacial area–relative permeability–saturation surfaces. They found that imbibition and drainage surfaces had an average difference of 7%. Nuske [2009] applied a pore-network model consisting of pore bodies only to obtain $p_c(S_w, a_{wn})$ relations for fractures. He discretized a fracture by a raster element model and interpreted each raster element as a pore body.

b. Lattice-Boltzmann methods.

Lattice-Boltzmann methods allow for the modeling of two-phase flow in a three-dimensional porous medium that is resolved in detail (e.g. using data from computer tomography scans). Unlike in pore-network models, it is easy to study a dynamic evolution of interfaces. However, unlike in pore-network models, the basis of Lattice-Boltzmann methods is not completely physical, but has been proven useful in practice. Using Lattice-Boltzmann schemes, $p_c(S_w, a_{wn})$ relationships (Porter et al. [2009]) and even $a_{ws}(S_w, p_c)$ and $a_{ns}(S_w, p_c)$ surfaces for "real" porous media have been calculated, see Ahrenholz et al. [2009].

5.7.2 Examples

5.7.2.1 Flow cell experiments

Niessner et al. [2008] developed a method to extract the relevant macro-scale parameters from pore-scale experiments. Therefore, they used results from Crandall et al. [2008] who developed a new method for creating experimental flow cells. These flow cells are created using stereolithography, i.e. by curing layers of photosensitive resins on the surface of a vat of heated resin with a laser. The flow cell geometry was created with a computer generated model, thus the geometry throughout the flow region is exactly known and can be used for data analysis. Experiments on air invading an initially water-filled flow cell have been performed. Knowing the exact pore geometry, matrix parameters such as porosity can be easily obtained. By recording a series of images as the non-wetting air invades the flow cell, data on dynamic evolution of interfaces, capillary pressure, saturation, interfacial areas and velocity of interfaces can be determined.

A crucial issue in this respect is the question of scales. While the experimental studies are performed on a pore scale and parameters are thus directly obtained on this scale, the numerical model is valid on a volume-averaged macro scale. Thus, it is

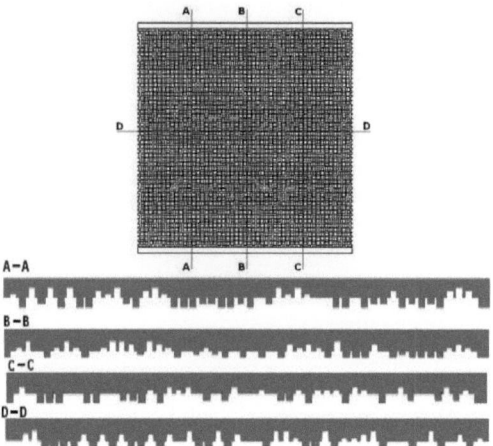

Figure 5.43: Flow cell and cross-sections of flow cell matrix. The heights of the cross-sections are magnified (taken from Crandall et al. [2008]).

necessary to determine a valid size of a representative elementary volume (REV) and average relevant parameters over these REVs. Laboratory scale experiments of two-phase flow in porous media have been conducted by numerous researchers, in various disciplines, and for a multitude of applications; a review by Buckley [1991] covers many of the models that have been used in these studies. For this current work, the porous medium used, referred to as the flow cell, has several unique aspects that enable the estimation of parameters required for the previously discussed, thermodynamically derived numerical model of two-phase flow in porous media. The greatest aid to the determination of parameters within this flow cell is that the entire geometry was created from a computer generated model, so that the pore-level structure of the porous medium is known.

The flow cell has a viewable flow area of $(10.16 \text{ cm})^2$, in which 5000 random sized throats are arranged in a lattice formation, see Fig. 5.43. The rectangular throat widths w_t vary from 0.35 mm to 1 mm while the throat heights h_t vary from 0.2 mm to 0.8 mm. These various h_t were assigned to the w_t in such a manner that the cross-sections of the throats are nearly square, with the smallest h_t relegated to the smallest w_t and vice versa as discussed in Crandall et al. [2008]. Additionally, the nearly square cross-sectional area of each throat allows for the estimation of the p_c within individual throats.

Experiments were conducted in the flow cell initially filled with water and placed horizontal (no gravitational forces in flow directions). Air was injected at a constant rate with a syringe pump while digital images were recorded with a CCD camera and stored on a computer for analysis. Additionally the pressure difference across the cell was measured with a pressure transducer. The pressure within the cell was observed to be less than 450 Pa above atmospheric pressure for even the highest flow rates studied here, so compressibility of the air is assumed to be negligible, see Crandall et al. [2008].

Experiments were run until no discernable change in the invading fluid structure was noted, after the air "brokethrough" the porous matrix into the exit manifold. Images were converted to binary data sets by first performing image analysis to isolate the invading air, then crop the image, and threshold the gray-scale picture afterwards. These images were then analyzed using an in-house code specific to this flow cell geometry to determine the volume of invaded fluid at the spatial location identified by the image. While the variation in h_t enabled a wider range of resistances to be studied, the additional code was required to determine the specific parameters required by the numerical model.

In order to determine the macro-scale parameters that are needed for the numerical model, the size of an averaging volume (REV size) needs to be determined. For this purpose, we subdivide the flow cell in n×n subvolumes and calculate porosity as a test parameter for all n×n subdomains. As the flowcell is constructed using a random distribution of w_t and h_t, the flow cell as a whole is to be considered as a homogeneous porous medium with a porosity of 35.79%. Note that this is the computational value of porosity resulting from an analysis of the computer generated flow cell geometry. The experimentally measured value of porosity is equal to 43 ± 2.5% and thus, much larger than the theoretical value. This might be due to slight variations from the production machine (over-curing may have led to a bit of shrinkage and to a rounding of corners).

In order to determine macro-scale parameters, the size of an averaging volume (REV size) needs to be determined. For this purpose, we subdivide the flow cell in n×n subvolumes and calculate porosity as a test parameter for all n×n subdomains. The subdivision of the flow cell for which porosity varies only little using the largest possible value for n determines the REV size. This is because, for all parameter analyses, the porous medium has to be considered as homogeneous.

We will proceed by showing how to determine the parameters of the numerical model: the static parameters S_w, p_c, and a_{wn} as well as the dynamic parameters $\underline{v}_{wn}$, $\underline{\underline{K}}_{wn}$, and E_{wn} that have not been determined so far.

Saturation, capillary pressure, and specific interfacial area. In order to obtain an interfacial area–capillary pressure–saturation surface, these three parameters need to be recorded during several drainage and imbibition cycles under static conditions. The experiments carried out in the flow cell have all taken place under dynamic conditions and only for primary drainage. Therefore, we only show how these parameters can in principle be obtained from flow cell data being aware of the fact that for the construction of an $a_{wn}(S_w, p_c)$ surface static data is actually needed. These experiments will be performed soon.

The calculation of saturation values in each of the n^2 REVs is straight forward. Identifying the pore throats being invaded by air within a specific REV i, summing up their total volume and dividing by the total pore volume of REV i allows for a direct calculation of the non-wetting phase (air) saturation S_n^i within REV i as

$$S_n^i = n^2 \cdot \frac{V_n^i}{V_t \cdot \phi^i}, \tag{5.129}$$

where V_n^i is the air volume within REV i, V_t is the total volume of the flow cell and ϕ is the porosity of REV i.

To obtain a macro-scale estimate of capillary pressure is equally simple. The location of the fluid–fluid interfaces are to be identified, local capillary pressure p_c^Γ at interface Γ within a pore throat can be calculated from the Young-Laplace equation

$$p_c = \sigma \left(\frac{2}{h_t} + \frac{2}{w_t} \right) \tag{5.130}$$

with σ equal to the interfacial tension between fluids and where the radii of the meniscus are assumed to be half of the h_t and w_t (Lenormand et al. [1983]). We are aware of the fact that this approximation is only a rough estimate. A more sophisticated approach is given by the Mayer-Princen-Stowe formula developed by Mayer and Stowe [1965] and Princen [1969]. The pore-scale capilary pressures can be averaged to obtain macro-scale capillary pressure p_c^i of REV i,

$$p_c^i = \frac{1}{I^i} \cdot \sum_{\Gamma=1}^{I^i} p_c^\Gamma, \tag{5.131}$$

where I^i is the number of fluid–fluid interfaces within REV i.

The third and last parameter needed in order to obtain an $a_{wn}(S_w, p_c)$ surface is specific interfacial area a_{wn}^i in each REV i. We assume that interfacial areas in the corners of the rectangular throats can be neglected, see Ma et al. [1996]. Thus, in a first approximation, the fluid–fluid interfacial area of interface Γ within a pore throat can be estimated as the procuct of its width and height

$$A_{wn}^\Gamma = h_t^\Gamma \cdot w_t^\Gamma. \tag{5.132}$$

Specific interfacial area a_{wn}^i in REV i can be calculated by summing up the local values of interfacial area A_{wn}^Γ of all interfaces Γ within REV i and dividing by the volume of the REV,

$$a_{wn}^i = \frac{n^2}{V_t} \cdot \sum_{\Gamma=1}^{I^i} \left(A_{wn}^\Gamma \right). \tag{5.133}$$

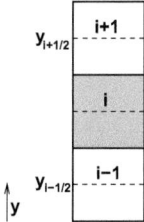

Figure 5.44: Notation used for the calculation of interfacial permeability K_{wn}.

Interfacial velocity and interfacial permeability. All remaining parameters of interest are dynamic parameters meaning that they can only be determined under dynamic conditions. To the best of our knowledge this is the first time that these interface-related dynamic parameters are estimated from physical experiments. In order to keep the equations for the determination of these parameter values as simple as possible we restrict ourselves to one-dimensional considerations (in y-direction). The extension to 2d is straightforward and is used for actually determining parameters from the flow cell experiments.

The first parameter of interest is interfacial velocity v_{wn}. It can be estimated by identifying the location of all interfaces Γ within REV i that are located in pore throats along the y-direction at two subsequent (very close) points in time. The locations of the interfaces at each of these two points in time is weighted by specific interfacial area and averaged over REV i and their difference is divided by the difference in time. The result is then divided by the sum of specific interfacial areas in REV i,

$$v_{wn}^i = \frac{1}{\sum_{I^i} A_{wn}^i} \frac{\sum_{\Gamma=1}^{I^i} y_1^\Gamma A_{wn,1} - \sum_{\Gamma=1}^{I^i} y_0^\Gamma A_{wn,0}}{t_1 - t_0}, \quad (5.134)$$

where the indices 1 and 0 denote the time levels, y is the location of the interface and t denotes time. Note that ideally, the two points in time are very close together so that the interfaces neither leave the pore throat they are located in nor enter or leave the REV.

The interfacial permeability tensor can be determined once interfacial velocity and interfacial area are known as these quantities are related as given in Eq. (5.4). Resolving this equation for the interfacial permeability component K_{wn} and stepping to a discrete description for REV i yields

$$K_{wn}^i = -\frac{v_{wn}^i}{\frac{a_{wn}^{i+\frac{1}{2}} - a_{wn}^{i-\frac{1}{2}}}{y_{i+\frac{1}{2}} - y_{i-\frac{1}{2}}}}. \quad (5.135)$$

This calculation is illustrated in Fig. 5.44. The solid lines correspond to REV boundaries. For the calculation of interfacial permeability K_{wn}^i of REV i, interfacial velocity in REV i as given in Eq. (5.134) is divided by the gradient in interfacial area. This gradient is calculated in discrete form by determining specific interfacial area with respect to the regions separated by dashed lines in Fig. 5.44 which are shifted from REV i by half an REV length in positive and negative y-direction.

Production rate of specific interfacial area. The production rate of specific interfacial area E_{wn}^i in REV i is given by Eq. (5.3) and consists of two parts, a storage part $E_{wn,I}^i$ and a flux part $E_{wn,II}^i$,

$$E_{wn}^i = E_{wn,I}^i + E_{wn,II}^i = \left.\frac{\partial a_{wn}}{\partial t}\right|_i + \nabla \cdot (a_{wn} v_{wn})|_i. \tag{5.136}$$

The first part $E_{wn,I}^i$ can be easily determined as

$$E_{wn,I}^i = \frac{a_{wn,1}^i - a_{wn,0}^i}{t_1 - t_0}. \tag{5.137}$$

The second part is more complicated. Writing it in completely discretized form yields the desired formula for the calculation of $E_{wn,II}^i$,

$$E_{wn,II}^i = \frac{a_{wn}^{i+\frac{1}{2}} v_{wn}^{i+\frac{1}{2}} - a_{wn}^{i-\frac{1}{2}} v_{wn}^{i-\frac{1}{2}}}{y_{i+\frac{1}{2}} - y_{i-\frac{1}{2}}}. \tag{5.138}$$

Results So far, only preliminary results could be obtained from the experiments. As only dynamic drainage experiments have been carried out, it is at the current stage not possible to construct a static $a_{wn}(S_w, p_c)$ surface. In order to do so, static $a_{wn}(S_w, p_c)$ data would be needed from several drainage and imbibition cycles. However, we show how these values can be obtained in principle. Therefore, we consider Fig. 5.45 where S_w, p_c, and a_{wn} are shown as a function of time in parts a), b), and c), respectively. If these were static experiments, we could pick $a_{wn}(S_w, p_c)$ data at selected pressure steps and use them to fit an $a_{wn}(S_w, p_c)$ surface. Note that in Fig. 5.45, specific interfacial area is always increasing. However, at some point within a drainage process, specific interfacial area decreases again. The strict increase of a_{wn} in Fig. 5.45 is due to the fact that in the experiments carried out so far, S_w has not yet decreased down to the point where destruction of interfaces takes place. This fact is visualized in part d) of Fig. 5.45 where a_{wn} is shown as a function of wetting-phase saturation. Extending this relation to lower S_w would lead to a maximum, and then a decrease in a_{wn}.

The experimental data has not been evaluated up to the point of calculating interfacial velocity, interfacial permeability and production rate of specific interfacial area

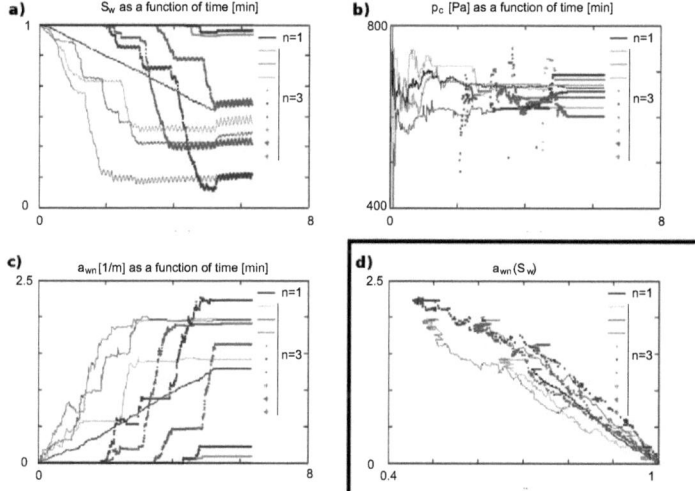

Figure 5.45: a) Wetting-phase saturation, b) capillary pressure, c) specific interfacial area as a function of time, and d) specific interfacial area as a function of wetting-phase saturation for $n = 3$.

Figure 5.46: Example of a three-dimensional pore network consisting of cylindrical pore throats and spherical pore bodies. *(Source: http://wwweng.uwyo.edu/chemical/research/piri/images/pore_scale.jpg)*

so far. However, for K_{wn}, first estimates showed that its value is in the range of $10^{-4} - 10^{-5} \left[\frac{m^3}{s}\right]$.

5.7.2.2 Pore-network models

A number of researchers used static pore network models in order to determine the specific interfacial area–capillary pressure–saturation relationship for two-phase flow in porous media. The principle behind is to first generate a network consisting of pore bodies and pore throats with a certain geometry and usually with a certain pore and throat size distribution. One example of such a network consisting of cylindrical pore throats and spherical pore bodies is shown in Fig. 5.46. Such a network is used in a second step to simulate static drainage and imbibition processes. Therefore, the network is usually assumed to be initially fully water-saturated. Then, the difference in phase pressure, $p_n - p_w$, at two opposing sides of the porous medium is increased. The value of $p_n - p_w$ is interpreted as capillary pressure. For each pressure step, the occupancy of the pores and throats is determined by applying the Young-Laplace equation and potentially specific trapping assumptions. From the occupancy of pore throats and bodies, saturation and interfacial areas are calculated based on the geometry and subsequently transferred to REV-based properties.

5.7.2.3 Lattice-Boltzmann simulations

Lattice-Boltzmann methods represent a tool to simulate dynamic behavior of fluids. Due to the involved computational effort, the method is usually restricted to

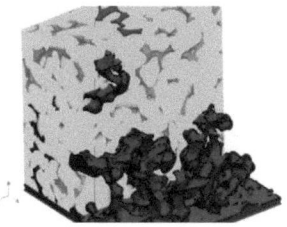

Figure 5.47: Identification of interfacial areas from results of a three-dimensional multiphase Lattice-Boltzmann simulation. *(Figure by Benjamin Ahrenholz, University of Braunschweig)*

small domains. Therefore, Lattice-Boltzmann methods are usually applied for representing pore-scale processes, although they are not automatically associated with a certain scale. Instead of solving the Navier-Stokes equations, the discrete Boltzmann equation is solved to simulate the flow of a Newtonian fluid and combined with a collision model. In Lattice-Boltzmann models, the considered fluids consist of a limited number of fictive particles which perform propagation and collision processes over a discrete lattice grid.

As Lattice-Boltzmann methods are dynamic methods, they are much more computationally costly than static pore network models. This is due to the fact that for each pressure step, the fluids will dynamically redistribute until the static configuration is reached. In theory, this takes infinitely long. Although this may seem a severe disadvantage, this fact is advantageous in two ways: first, when performing physical experiments, the dynamic redistribution of fluids after a pressure step represents a physical real-life process: in practice, it is impossible to reach the static condition at once. Second, unlike static pore network models, Lattice-Boltzmann methods allow for the modeling of the movement of fluid phases through a porous medium which is dynamic by nature. Furthermore, they allow for the inclusion of complex (realistic) geometries obtained e.g. from a CT scan, see e.g. Fig. 5.47. Using Lattice-Boltzmann simulations, Ahrenholz et al. [2009] obtained $a_{wn}(S_w, p_c)$ relationships. They additionally calculated $a_{ws}(S_w, p_c)$ and $a_{ns}(S_w, p_c)$ surfaces, see Sec. 3.5.

5.8 Summary

In this chapter, we have modeled various systems where the interfacial-area-based approach provides advantages over the classical approach. These advantages are either due to the fact that both drainage and imbibition take place (capillary hysteresis will occur), or due to kinetic mass and / or energy transfer which cannot be described in a physically correct way using the classical model. Specifically, we considered

- immiscible isothermal two-phase flow (drainage only)
- horizontal capillary redistribution (drainage and imbibition)
- kinetic interphase mass transfer
- kinetic interphase mass and energy transfer
- heterogeneous media and physically correct treatment of material interfaces (drainage and imbibition)
- gas–water processes in fracture–matrix systems (both kinetic interphase mass transfer as well as drainage and imbibition).

We also discussed how important parameters and constitutive relationships of the interfacial-area-based model can be obtained experimentally or numerically. Despite of the advantages of the interfacial-area-based approach in connection with the above applications concerning the physical basis of the two-phase flow description it has to be stressed that the interfacial-area-based approach is computationally more expensive than the classical approach. This is due to the number of balance equations that need to be solved and to the complexity of the constitutive equations that need to be evaluated. Therefore, a strategy needs to be found in order to reduce the computing time. An idea is to identify domains or time periods where the computationally more expensive interfacial-area-based model needs to be solved. In all other domains or time periods, it is sufficient to use the computationally less expensive, but not always physically based classical model. The purpose of the next chapter is to find a strategy to meet these needs.

6 Multi-scale–multi-physics modeling: a perspective

Having realized the dilemma between applying a model that is able to account for hysteresis and kinetic interphase mass / energy transfer in a physically based way on the one hand and computational efficiency on the other hand, a strategy to find a good compromise between physical correctness and numerical efficiency needs to be found. While the common approach for solving processes occurring in a domain is to select the maximum complexity of these processes and the smallest relevant scale for the numerical solution of the associated processes, multi-scale multi-physics approaches optimize this choice: processes are solved for only in relevant subdomains and on relevant scales. This means on the one hand that the most complex processes are not generally solved for in the total domain, but only in a subdomain where they are relevant. In the rest of the domain, simpler models may be applied. On the other hand, only processes which are highly dependent on small-scale effects are resolved on a small scale. Other processes may be resolved on larger scales. There are two main advantages of multi-scale multi-physics approaches:

1. a reduction of the amount of necessary data (may have a significant economic impact), and
2. a reduction of computing times, thus making the solution of very large problems possible that might otherwise not be numerically treated.

Fig. 6.1 shows a general physical system which typically consists of certain subregions. The resolved regions represent subdomains of the total system in which other physical processes than in the remaining domain dominate. These processes take

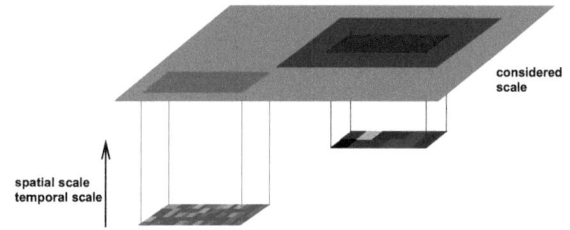

Figure 6.1: Schematic representation of spatial and temporal scales.

place on finer spatial and temporal scales. While traditional models describe this system on one scale—this must be the fine scale if an accurate solution is desired—**multi-scale algorithms** take the dependence of the processes on both spatial and temporal scales into account. In the frame of two-phase flow modeling in porous media, much research has been done in the last decades to upscale either pressure or saturation equation (Durlofsky [1991], Renard and de Marsily [1997], Efendiev et al. [2000], Efendiev and Durlofsky [2002], Chen et al. [2003]) or include the different scales directly in the numerical scheme by using multi-scale finite volumes or elements, see e.g. E and Engquist [2003a], E et al. [2003], Chen and Hou [2002], Hou and Wu [1997], Jenny et al. [2003]. **Multi-physics algorithms**—in constrast to the abilities of traditional models—allow to apply different model concepts in different subdomains. In this respect, research has advanced in the context of domain decomposition techniques (see e.g. Wheeler et al. [1999], Yotov [2002]) and in the context of mortar finite element techniques that allow multi-physics as well as multi-numerics coupling (see e.g. Peszynska et al. [2002]).

Niessner and Helmig [2006, 2007, 2009] combined both multi-physics and multi-scale approaches to a multi-scale multi-physics framework. They used up- and downscaling techniques (vertical coupling) as well as coupling of different physical processes on one scale (horizontal coupling) and showed that the created algorithm is capable of reducing computational demands.

In the following, a more detailed overview of current multi-scale, multi-physics and combined multi-scale multi-physics approaches is given (Sec. 6.1). In Sec. 6.2, it is discussed how dimensional analysis can help to integrate the interfacial-area-based approach into this multi-scale multi-physics framework. This procedure is illustrated using the example of carbon dioxide storage in Sec. 6.3 and in Sec. 6.4, a chapter summary is provided.

6.1 Overview of multi-scale multi-physics approaches

In multi-scale modeling, more than one scale is involved in the modeling as the name implies. In general, these two or more scales are coupled in a multi-directional way. This means, it is not sufficient to provide only upscaling (or only downscaling) operators. Instead, both kinds of operators are generally needed. Only special applications with weak coupling between the scales allow for a mono-directional coupling and thus, only upscaling or only downscaling operators.

Classical upscaling strategies comprise rigorous methods like the method of asymptotic expansions (homogenization) and volume averaging. These methods allow to rigorously derive coarse-scale equations from known fine-scale equations if certain assumptions can be made. The most crucial issues in connection to the method of asymptotic expansions are the assumptions of periodicity and scale separation.

Scale separation means that the characteristic length l on the small scale is related to the characteristic length L on the large scale by

$$\epsilon = \frac{l}{L}, \qquad (6.1)$$

where ϵ is very close to zero. For volume averaging, in all but the very simple cases closure relations need to be found for products of deviation quantities which live on the small scale and do not disappear in the averaging process. Thus, the challenge is the closure of the coarse-scale equation system that allows to capture the influence of fine-scale fluctuations on coarse-scale properties. More heuristic methods comprise the method of effective coefficients where a coarse-scale equation is postulated and coefficients of this equation (effective parameters) are sought by solving local fine-scale single-phase problems (downscaling). A very similar method is the method of pseudo functions, where instead of parameters, macro-scale relationships (relative permeability, mobility, or fractional flow funcions) are obtained by solving local fine-scale multi-phase problems. For downscaling, the typical methodology is to specify boundary conditions at the boundaries of a coarse-grid block and solve a fine-grid problem in the respective domain. The boundary conditions are obtained either directly from the coarse-scale problem or coarse-scale results are re-scaled to fine-scale properties using fine-scale material parameters. In the latter case, fine-grid boundary conditions can be specified along the boundaries of the downscaling domain.

It is also possible to directly include the knowledge about the scales in the numerical scheme. This idea has been put into practice in the frame of the homogeneous and heterogeneous multi-scale method (E et al. [2007], E. and Engquist [2003b], E and Engquist [2003a]), the variational multi-scale method (Larson and Målqvist [2009]), multiscale finite volumes (Chen and Hou [2002], Jenny et al. [2003]), and multi-scale finite elements (Larson and Målqvist [2009], Hou and Wu [1997]). Nordbotten and Bjørstad [2008] compared multi-scale finite volumes to domain decomposition and found that the multi-scale finite volume method is a special case of a nonoverlapping domain decomposition preconditioner.

Multi-physics strategies refer to models where different physical processes are solved in one model domain. One can distinguish horizontal coupling where different model concepts are solved on the same scale in different subdomains and vertical coupling where only one model concept is applied, but on different scales. **Vertical** coupling has been widely used in atmospheric and ocean modeling, where usually a fine-scale model is nested into a coarse-scale model. Here, the boundary conditions for the fine-scale model represent the crucial issue. Koch [1987], e.g. gives an overview of these nested techniques for weather-forecasting models. In these models, the coarse-scale problem influences the fine-scale problem, but generally, there is no transfer of information from the fine-scale model to the coarse-scale model.

Horizonal coupling has been developed for multi-phase issues by Wheeler et al. [1999] and Yotov [2002], who subdivided their system into a number of non-overlapping subdomains according to the relevance of physical processes. However, the issue of coupling these different models in a physically and mathematically correct way still represents a crucial issue. **Multi-physics algorithms** as horizontal coupling algorithms allow to apply different model concepts in different subdomains. In this respect, research has advanced in the context of domain decomposition techniques (see e.g. Wheeler et al. [1999], Yotov [2002]) and in the context of mortar finite element techniques that allow multi-physics as well as multi-numerics coupling (see e.g. Peszynska et al. [2002]).

Recently, Niessner and Helmig [2006, 2007, 2009] combined multi-scale and multi-physics modeling to come up with a multi-scale multi-physics approach which allows to couple processes both on different scales and in different subdomains. For horizontal domains without capillary pressure effects, Niessner and Helmig [2006] coupled a subdomain two-phase two-component model on a fine scale to a two-phase model on a coarse scale occurring in the complete domain. They solved a hyperbolic concentration equation explicitly in the subdomain using micro time steps, the saturation equation in an upscaled and time-explicit form on a coarse grid and the pressure equation implicitly on a fine scale and grid. In Niessner and Helmig [2007], this approach was extended to describe the coupling of local three-phase three-component processes to a global two-phase flow problem. This means that an additional concentration equation was solved. Niessner and Helmig [2009] upscaled also the pressure equation to the coarse scale such that their approach can be interpreted as a downscaling approach where a global problem is solved on a coarse scale and in highly active regions, a more complex problem is solved on a fine scale. In their algorithm, a reconstruction of the fine-scale velocity field became necessary which was put into practice by means of a local reconstruction method proposed in Chen et al. [2003].

This multi-scale multi-physics framework is currently being extended at the Department of Hydromechanics and Modeling of Hydrosystems, see Helmig et al. [2010]. Fritz et al. [2009] have implemented a multi-physics coupling for a global single-phase two-component model to a local two-phase two-component model for non-isothermal compressible flow including gravity effects. They also developed a strategy to extend or shrink the local problem depending on phase appearance / disappearance in an adaptive way. The framework is being further extended by Markus Wolff who includes both capillary pressure and gravity into a multi-scale framework. He calculates both an advective and a dispersive correction to the macro-scale saturation equation based on the solution of local or extended local problems over coarse-grid blocks.

6.2 Multi-scale multi-physics based on dimensional analysis

As argued in Sec. 5.8, the solution of the coupled system of nonlinear partial differential equations in the frame of the alternative approach is very challenging and time consuming, especially if kinetic interphase mass and / or energy transfer are considered. This is mainly due to the fact that more partial differential equations need to be solved. If, for example, a two-phase system is considered, two partial differential equations need to be solved when using the classical approch while three partial differential equations need solving in the frame of the alternative approach. Similarly, for two-phase two-component flow and transport in porous media, the classical approach calls for the solution of two partial differential equations while the alternative approach includes three additional partial differential equations. This creates a dilemma between physical correctness and computational speed: the alternative approach includes more physics, but is computationally more expensive. This is where multi-scale multi-physics approaches offer extreme advantages: the best approach can be chosen as a function of the governing physical processes. If, e.g. in some part of the domain of interest or over a certain time interval hysteresis is not important (for example, a drainage process in a homogeneous porous medium) the computationally cheaper classical approach can be used; if, contrarily, hysteresis occurs, the alternative approach must be used. Similar considerations can be made with respect to kinetic interphase mass and / or energy transfer. This calls for physically based indicators which allow to choose the "best" model. Dimensionless numbers are an ideal option for such indicators. In the following, appropriate choices of dimensionless numbers / physically based indicators are discussed with respect to the three key issues necessitating use of the alternative model: hysteresis, kinetic interphase mass transfer, and kinetic interphase energy transfer.

6.2.1 Hysteresis

Hysteresis in two-phase flow in porous media generally refers to the fact that there is a non-single-valued relationship between capillary pressure and saturation. Hysteretic effects occur when both drainage and imbibition take place within the considered domain or / and within the time frame of interest. In that case the sign of the time rate of change of saturation, $\frac{\partial S_w}{\partial t}$, changes in time, in space, or both in space and time. If, contrarily, there is no change in the sign of $\frac{\partial S_w}{\partial t}$ in time or space, a single-valued capillary pressure–saturation relationship can be specified. In the former case (hysteresis occurs) the alternative approach provides advantages over the classical approach as it includes hysteresis in a natural way without necessitating the specification of myriads of scanning curves. In the latter case, the classical approach gives equally good results as the alternative approach, but needs one partial differential equation less and is computationally cheaper in consequence.

These considerations lead to a strategy that allows to decide when the computationally more expensive alternative approach needs to be used and when the computationally cheaper classical approach is sufficient: if $\frac{\partial S_w}{\partial t}(\underline{x}, t^*)$ has the same sign for all values of $\underline{x}$ at a point t^* in time and equals the sign of $\frac{\partial S_w}{\partial t}(\underline{x}, t)$ for all $t < t^*$ the classical model may be used. In all other cases, only the alternative approach will give physically based results.

6.2.2 Kinetic interphase mass transfer

Depending on the parameters, initial conditions, and boundary conditions of the system, kinetics might be more or less important for mass transfer. To investigate this importance, the system of equations (5.69) through (5.81) is made dimensionless and the dependence of kinetics on Damköhler number and Peclet number is studied, see Niessner and Hassanizadeh [2009a].

To do so, dimensionless time t^*, nabla operator ∇^*, phase velocity $\underline{v}_\alpha^*$, external source $Q_\alpha^{\kappa*}$, specific interfacial area a_{wn}^*, dispersion coefficient $\bar{D}_\alpha^{\kappa*}$, interfacial velocity $\underline{v}_{wn}^*$, interfacial production term E_{wn}^*, phase gravity $\underline{g}_\alpha^*$, interfacial permeability K_{wn}^*, phase pressure p_α^*, capillary pressure p_c^*, density ratio ρ_n^*, and viscosity ratio μ_w^* are defined by

$$t^* = \frac{t v_R}{\phi L}, \qquad \nabla^* = L \nabla, \qquad \bar{v}_\alpha^* = \frac{\bar{v}_\alpha}{v_R}, \qquad Q_\alpha^{\kappa*} = \frac{Q_\alpha^\kappa L}{X_{\alpha,s}^\kappa v_R}, \tag{6.2}$$

$$a_{wn}^* = \frac{a_{wn}}{a_{R,wn}}, \qquad \bar{D}_\alpha^{\kappa*} = \frac{\bar{D}_\alpha^\kappa}{D_{R,\alpha}}, \qquad \underline{v}_{wn}^* = \frac{\phi}{v_R} \underline{v}_{wn}, \tag{6.3}$$

$$E_{wn}^* = \frac{a_{R,wn} \phi L}{v_R} E_{wn}, \qquad \underline{g}_\alpha^* = \frac{\bar{\rho}_\alpha g L}{p_R}, \qquad K_{wn}^* = \frac{\phi K_{wn}}{L a_{R,wn} v_R}, \tag{6.4}$$

$$p_\alpha^* = \frac{p_\alpha}{p_R}, \qquad p_c^* = \frac{p_c}{p_R}, \qquad \rho_n^* = \frac{\bar{\rho}_n}{\bar{\rho}_w}, \qquad \mu_w^* = \frac{\mu_w}{\mu_n}, \tag{6.5}$$

where v_R is a reference velocity, L is a characteristic length, $a_{R,wn}$ is a reference specific interfacial area, $D_{R,\alpha}$ is reference dispersion coefficient, and p_R is a reference pressure. It is assumed that p_R and v_R can be chosen such that $\frac{K p_R}{\mu_w v_R L} = 1$. Also, Peclet number Pe_α and Damköhler number Da^κ are defined by

$$\text{Pe}_\alpha = \frac{v_R L}{D_{R,\alpha}}, \qquad \text{Da}^\kappa = \frac{D^\kappa L a_{R,wn}}{d v_R}. \tag{6.6}$$

These definitions lead to the following dimensionless form of Eq. (5.69) through (5.81):

$$\frac{\partial}{\partial t^*}\left(S_w \bar{X}_w^w\right) + \nabla^* \cdot \left(\bar{X}_w^w \underline{v}_w^*\right) - \nabla^* \cdot \left(\frac{D_w^{w*}}{\mathrm{Pe}_w}\nabla^* \bar{X}_w^w\right) = Q_w^{w*} - \mathrm{Da}^w a_{wn}^* \rho_n^* \left(X_{n,s}^w - \bar{X}_n^w\right)$$

$$\frac{\partial}{\partial t^*}\left(S_w \bar{X}_w^a\right) + \nabla^* \cdot \left(\bar{X}_w^a \underline{v}_w^*\right) - \nabla^* \cdot \left(\frac{D_w^{a*}}{\mathrm{Pe}_w}\nabla^* \bar{X}_w^a\right) = Q_w^{a*} + \mathrm{Da}^a a_{wn}^* \left(X_{w,s}^a - \bar{X}_w^a\right)$$

$$\frac{\partial}{\partial t^*}\left(S_n \bar{X}_n^w\right) + \nabla^* \cdot \left(\bar{X}_n^w \underline{v}_n^*\right) - \nabla^* \cdot \left(\frac{D_n^{w*}}{\mathrm{Pe}_n}\nabla^* \bar{X}_n^w\right) = Q_n^{w*} + \mathrm{Da}^w a_{wn}^* \left(X_{n,s}^w - \bar{X}_n^w\right)$$

$$\frac{\partial}{\partial t^*}\left(S_n \bar{X}_n^a\right) + \nabla^* \cdot \left(\bar{X}_n^a \underline{v}_n^*\right) - \nabla^* \cdot \left(\frac{D_n^{a*}}{\mathrm{Pe}_n}\nabla^* \bar{X}_n^a\right) = Q_n^{a*} - \mathrm{Da}^a a_{wn}^* \frac{1}{\rho_n^*} \left(X_{w,s}^a - \bar{X}_w^a\right)$$

$$\frac{\partial a_{wn}^*}{\partial t^*} + \nabla^* \cdot (a_{wn}^* \underline{v}_{wn}^*) = E_{wn}^*$$

$$\underline{v}_w^* = -S_w^2 \left(\nabla^* p_w^* - \underline{g}_w^*\right)$$

$$\underline{v}_n^* = -S_n^2 \mu_w^* \left(\nabla^* p_n^* - \underline{g}_n^*\right)$$

$$\underline{v}_{wn}^* = -K_{wn}^* \nabla^* a_{wn}^*$$

$$p_c^* = p_n^* - p_w^*$$

$$S_w + S_n = 1$$

$$\bar{X}_w^w + \bar{X}_w^a = 1$$

$$\bar{X}_n^w + \bar{X}_n^a = 1$$

$$a_{wn}^* = a_{wn}^* \left(S_w, p_c^*\right).$$

In order to investigate the importance of kinetics, we define $\mathrm{Pe} = \mathrm{Pe}_w = \mathrm{Pe}_n$ and $\mathrm{Da} = \mathrm{Da}^w = \mathrm{Da}^a$ and vary Pe and Da independently by taking one and two orders of magnitude larger and smaller numbers than Damköhler Da_0 and Peclet number Pe_0 of the dimensional numerical test case of Sec. 5.3.3. In order to vary Pe, different values of $\bar{D}_\alpha^\kappa$ are used, while for the variation of Da, the diffusion length d^κ is varied in the numerical model and the setup of the dimensional example is maintained.

Fig. 6.2 shows a comparison of actual mass fractions $\bar{X}_w^a$ and $\bar{X}_n^w$ to solubility mass fractions $X_{s,w}^a$ and $X_{s,n}^w$ for five different Damköhler numbers. Therefore, a cut through the domain at $y = 0.25$ m is shown and two different time steps are compared.

It is clear that the system is practically instantaneously in equilibrium with respect to the mass fraction $\bar{X}_n^w$ (water mass fraction in the gas phase) for the whole range of considered Damköhler numbers. With respect to the mass fraction $\bar{X}_w^a$ (air mass fraction in the water phase), for low Damköhler numbers and early times, the system is far from equilibrium. With increasing time and with increasing Damköhler

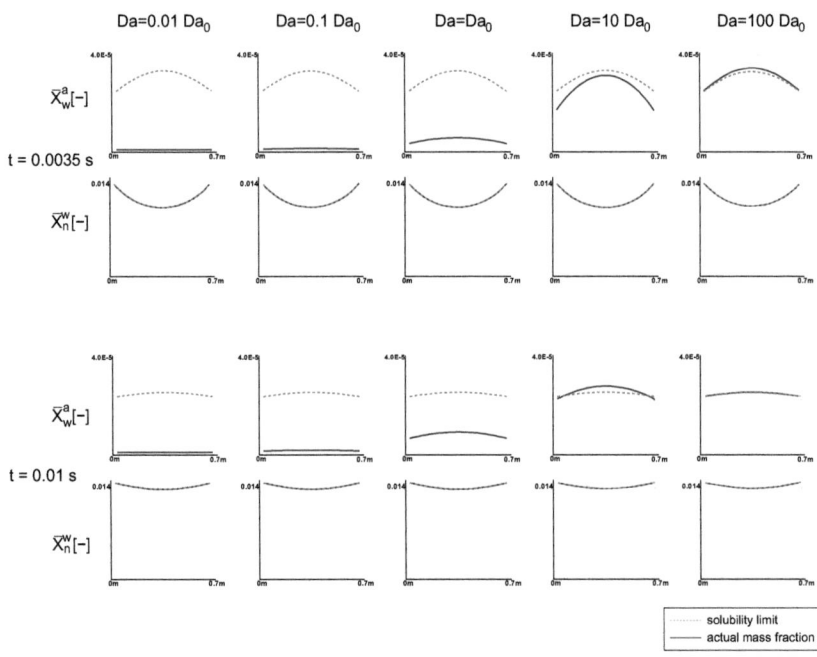

Figure 6.2: Solubility limits $X_{s,w}^a$ and $X_{s,n}^w$ and actual mass fractions $\bar{X}_w^a$ and $\bar{X}_n^w$ along the line $y = 0.25$ m for two different time steps and five different Damköhler numbers (Da_0 is the Damköhler number of the dimensional example of Sec. 5.3.3).

number, the system approaches equilibrium. As for high Damköhler numbers mass transfer is very fast, an "overshoot" occurs and the system becomes oversaturated before it reaches equilibrium.

One might argue that the considered time steps are extremely small and not relevant for the time scale relevant for the whole domain. However, what happens on this very small time has a high influence on the state of the system at all subsequent times as was also indicated by the dimensional example.

It turned out that for different Peclet numbers, there is no difference in results. That means that kinetic interphase mass transfer is independent of Peclet number, at least within the four orders of magnitude considered here.

6.2.3 Kinetic interphase energy transfer

In the following, non-isothermal two-phase flow with interphase heat and mass transfer in a rigid, isotropic, and incompressible porous medium is considered. The areal mass density of the wn-interface, Γ_{wn}, is assumed to be constant which seems a reasonable assumption since no surfactants are involved that would accumulate on the interface. Fluid densities and viscosities are considered pressure- and temperature-dependent.

Starting from the equation system given by Eq.s (5.86) through (5.109), Niessner and Hassanizadeh [2009c] proceed by defining the following dimensionless variables:

$$t^* = \frac{t v_R}{\phi L}, \quad \nabla^* = L \nabla, \quad \underline{v}_\alpha^* = \frac{\underline{v}_\alpha}{v_R}, \quad Q_\alpha^{\kappa *} = \frac{Q_\alpha^\kappa L}{v_R}, \quad a_{\alpha\beta}^* = \frac{a_{\alpha\beta}}{a_R}, \quad \rho_\alpha^* = \frac{\rho_\alpha}{\rho_{R,\alpha}},$$

$$\underline{\underline{D}}_\alpha^{\kappa *} = \frac{\underline{\underline{D}}_\alpha^\kappa}{D_{R,\alpha}}, \quad \underline{v}_{wn}^* = \frac{\underline{v}_{wn}}{v_R}, \quad \underline{E}_{wn}^* = \frac{\phi L}{v_R a_R} \underline{E}_{wn}, \quad \underline{g}^* = \frac{\rho_{R,\alpha} \underline{g} L}{p_R}, \quad T_\alpha^* = \frac{T_\alpha}{T_R},$$

$$p_\alpha^* = \frac{p_\alpha}{p_R}, \quad p_c^* = \frac{p_c}{p_R}, \quad \rho^* = \frac{\rho_{R,n}}{\rho_{R,w}}, \quad R_{\alpha,Q} = \frac{\rho_{\alpha,Q}}{\rho_{R,w}}, \quad \mu_\alpha^* = \frac{\mu_\alpha}{\mu_{R,\alpha}}, \quad \mu^* = \frac{\mu_{R,w}}{\mu_{R,n}},$$

$$h_\alpha^* = \frac{h_\alpha}{h_R}, \quad h_{\alpha,Q}^* = \frac{h_{\alpha,Q}}{h_R}, \quad C_{p,wn}^* = \frac{T_R}{h_R} C_{p,wn}, \quad C_{p,w}^* = \frac{T_R}{h_R} C_{p,w}, \quad e_{wn}^* = \frac{e_{wn}}{a_R},$$

$$C_{p,n}^* = \frac{T_R}{h_R} C_{p,n}, \quad C_{p,s}^* = \frac{T_R(1-\phi)}{h_R \phi} C_{p,s}, \quad \underline{\underline{D}}^{th*} = \frac{\underline{\underline{D}}_\alpha^{th}}{\lambda_R}, \quad \lambda_\alpha^* = \frac{\lambda_\alpha}{\lambda_R}, \quad \lambda_{\alpha\beta}^* = \frac{\lambda_{\alpha\beta}}{\lambda_R},$$

where v_R is a reference velocity, L is a characteristic length, a_R is a reference specific interfacial area, $\rho_{R,\alpha}$ a reference density and $\mu_{R,\alpha}$ a reference viscosity of phase α, $D_{R,\alpha}$ is a reference dispersion coefficient, and p_R is a reference pressure. Furthermore, T_R is a reference temperature, h_R a reference specific enthalpy, and λ_R a reference thermal conductivity. It is assumed that p_R and v_R can be chosen such that $\frac{K p_R}{\mu_{R,w} v_R L} = 1$ and a_R and v_R such that $\frac{a_R K_{wn}}{L v_R} = 1$. Also, Peclet number Pe_α, Damköhler number Da^κ, and Nusselt number $Nu_{\alpha\beta}$ are defined by

$$Pe_\alpha = \frac{v_R L}{D_{R,\alpha}}, \qquad Da^\kappa = \frac{D^\kappa L a_R}{d v_R}, \qquad Nu_{\alpha\beta} = \frac{\rho_{R,\alpha} h_R v_R d_{\alpha\beta}^T}{\lambda_R T_R L a_R}. \tag{6.7}$$

Furthermore, mass ratio and thermal diffusion length of the α-phase are given by

$$M_\alpha = \frac{\Gamma_{wn} a_R}{\rho_{R,\alpha} \phi S_\alpha}, \qquad L_\alpha^{th} = \frac{\lambda_R T_R}{\rho_{R,\alpha} h_R v_R}. \tag{6.8}$$

The mass ratio is the ratio of mass within the wn-interface over the mass within the bulk α-phase within an REV.

These definitions lead to the following dimensionless form of Eq. (5.86) through (5.109):

$$\frac{\partial}{\partial t^*}\left(\rho_w^* S_w X_w^w\right) + \nabla^* \cdot \left(\rho_w^* X_w^w \underline{v}_w^*\right) - \nabla^* \cdot \left(\frac{D_w^{w*}}{\text{Pe}_w}\nabla^* \rho_w^* X_w^w\right)$$
$$= R_{w,Q} Q_w^{w*} - \text{Da}^w a_{wn}^* \rho_w^* \left(X_{n,s}^w - X_n^w\right) \quad (6.9)$$

$$\frac{\partial}{\partial t^*}\left(\rho_w^* S_w X_w^a\right) + \nabla^* \cdot \left(\rho_w^* X_w^a \underline{v}_w^*\right) - \nabla^* \cdot \left(\frac{D_w^{a*}}{\text{Pe}_w}\nabla^* \rho_w^* X_w^a\right)$$
$$= R_{w,Q} Q_w^{a*} + \text{Da}^a a_{wn}^* \rho_w^* \left(X_{w,s}^a - X_w^a\right) \quad (6.10)$$

$$\frac{\partial}{\partial t^*}\left(\rho_n^* S_n X_n^w\right) + \nabla^* \cdot \left(\rho_n^* X_n^w \underline{v}_n^*\right) - \nabla^* \cdot \left(\frac{D_n^{w*}}{\text{Pe}_n}\nabla^* \rho_n^* X_n^w\right)$$
$$= R_{n,Q} Q_n^{w*} + \text{Da}^w a_{wn}^* \rho_n^* \left(X_{n,s}^w - X_n^w\right) \quad (6.11)$$

$$\frac{\partial}{\partial t^*}\left(\rho_n^* S_n X_n^a\right) + \nabla^* \cdot \left(\rho_n^* X_n^a \underline{v}_n^*\right) - \nabla^* \cdot \left(\frac{D_n^{a*}}{\text{Pe}_n}\nabla^* \rho_n^* X_n^a\right)$$
$$= R_{n,Q} Q_n^{a*} - \text{Da}^a a_{wn}^* \frac{1}{\rho^*}\rho_n^* \left(X_{w,s}^a - X_w^a\right) \quad (6.12)$$

$$\frac{\partial a_{wn}^*}{\partial t^*} + \nabla^* \cdot \left(a_{wn}^* \underline{v}_{wn}\right) = E_{wn}^* \quad (6.13)$$

$$\frac{\partial \left(\rho_w^* S_w h_w^*\right)}{\partial t^*} + \phi \nabla^* \cdot \left(\rho_w^* S_w h_w^* \underline{v}_w^*\right) + 0.25 \cdot S_w M_w C_{p,wn}^* \left(T_n^* - T_w^*\right) E_{wn}^*$$
$$= \frac{L_w^{th}}{L}\nabla^* \cdot \left(\underline{\underline{D}}_w^{th*}\nabla^* T_w^*\right) + \frac{1}{\text{Nu}_{wn}}a_{wn}^* \lambda_{wn}^* \left(T_n^* - T_w^*\right)$$
$$+ \frac{1}{\text{Nu}_{ws}}a_{ws}^* \lambda_{ws}^* \left(T_s^* - T_w^*\right)$$
$$+ R_{w,Q} h_{w,Q}^* \left(Q_w^{w*} + Q_w^{a*}\right) \quad (6.14)$$

$$\frac{\partial \left(\rho_n^* S_n h_n^*\right)}{\partial t^*} + \phi \nabla^* \cdot \left(\rho_n^* S_n h_n^* \underline{v}_n^*\right) + 0.25 \cdot S_n M_n C_{p,wn}^* \left(T_w^* - T_n^*\right) E_{wn}^*$$
$$= \frac{L_n^{th}}{L}\nabla^* \cdot \left(\underline{\underline{D}}_n^{th*}\nabla^* T_n^*\right) - \frac{1}{\text{Nu}_{nw}}a_{wn}^* \lambda_{wn}^* \left(T_n^* - T_w^*\right)$$
$$+ \frac{1}{\text{Nu}_{ns}}a_{ns}^* \lambda_{ns}^* \left(T_s^* - T_n^*\right)$$
$$+ R_{n,Q} h_{n,Q}^* \left(Q_n^{w*} + Q_n^{a*}\right) \quad (6.15)$$

$$C_{p,s}^* \frac{\partial T_s^*}{\partial t^*} = \frac{L_s^{th}}{L}\nabla^* \cdot \left(\underline{\underline{D}}_s^{th*}\nabla^* T_s^*\right) - \frac{1}{\text{Nu}_{sw}}a_{ws}^* \lambda_{ws}^* \left(T_s^* - T_w^*\right)$$
$$- \frac{1}{\text{Nu}_{sn}}a_{ns}^* \lambda_{ns}^* \left(T_s^* - T_n^*\right) \quad (6.16)$$

$$\underline{v}_w^* = -S_w^2 \mu_w^* \left(\nabla^* p_w^* - \rho_w^* \underline{g}^*\right) \quad (6.17)$$

$$\underline{v}_n^* = -S_n^2 \mu_n^* \left(\nabla^* p_n^* - \rho_n^* \underline{g}_n^*\right) \quad (6.18)$$

$$\underline{v}_{wn}^* = -\nabla^* a_{wn}^* \quad (6.19)$$

$$\underline{\underline{D}}_w^{th*} = \lambda_w^* \underline{\underline{I}} + \frac{\alpha_T}{L_w^{th}} C_{p,w}^* |\underline{v}_w^*| \underline{\underline{I}} + \frac{\alpha_L - \alpha_T}{L_w^{th}} C_{p,w}^* \frac{\underline{v}_w^* \otimes \underline{v}_w^*}{|\underline{v}_w^*|} \quad (6.20)$$

$$\underline{\underline{D}}_n^{th*} = \lambda_n^* \underline{\underline{I}} + \frac{\alpha_T}{L_n^{th}} C_{p,n}^* |\underline{v}_n^*| \underline{\underline{I}} + \frac{\alpha_L - \alpha_T}{L_n^{th}} C_{p,n}^* \frac{\underline{v}_n^* \otimes \underline{v}_n^*}{|\underline{v}_n^*|} \quad (6.21)$$

$$\underline{\underline{D}}_s^{th*} = \lambda_s^* \underline{\underline{I}} \quad (6.22)$$

$$p_c^* = p_n^* - p_w^* \quad (6.23)$$

$$S_w + S_n = 1 \quad (6.24)$$

$$X_w^w + X_w^a = 1 \quad (6.25)$$

$$X_n^w + X_n^a = 1 \quad (6.26)$$

$$a_{wn}^* = a_{wn}^* (S_w, p_c^*) \quad (6.27)$$

$$E_{wn}^* = -e_{wn}^* \frac{\partial S_w}{\partial t^*}. \quad (6.28)$$

Having obtained this complex dimensionless equation system, it is possible to identify limiting situations that admit certain simplifications.

- Kinetics of mass transfer has to be taken into account if Damköhler numbers are low; if they are high, a standard multi-phase multi-component model assuming equilibrium within each representative elementary volume, with mass fractions given by Henry's and Raoult's laws, is reasonable.

- If Peclet numbers are high (or Damköhler numbers are low), the diffusive / dispersive terms in the mass balance equations may be neglected.

- Typically, the mass ratios M_w and M_n are small. For an air–water system, for example, $M_w \approx 10^{-8}$ and $M_n \approx 10^{-5}$. In that case, the last term on the left hand sides of the energy balance equations for wetting and non-wetting phase could be neglected.

- The first term on the right hand side of wetting- and non-wetting phase energy balance equations could only be neglected, if the characteristic length L is very large, e.g. if the modeling domain is much larger than the characteristic thermal lengths which are in the order of $L_\alpha^{th} \approx 1$ to 10.

- If Nusselt numbers $\mathrm{Nu}_{\alpha\beta}$ are small, thermal equilibrium within a representative elementary volume can be assumed meaning that all three phases have the same temperature. As a matter of fact, these Nusselt numbers are most often very small so that for many situations, the assumption of local thermal equilibrium is reasonable. But if flow velocity is high, if characteristic length is small, if thermal diffusion length is large, or if specific interfacial area is small, thermal non-equilibrium situations will occur.

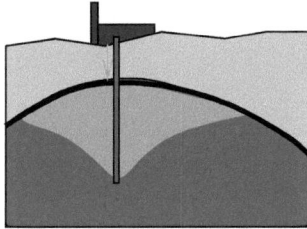

Figure 6.3: Carbon dioxide injection into the subsurface.

6.3 An example: carbon dioxide storage in the subsurface

In order to illustrate the application of characteristic parameters and dimensionless numbers with respect to multi-physics methods for the coupling of interfacial-area-based and classical two-phase flow model the example of CO_2 storage in the subsurface will be considered. Fig. 6.3 recalls the situation: in order to mitigate the greenhouse effect, CO_2 is injected into the subsurface, often below a dome-shaped impermeable cap rock. After some time, the injection is stopped. Advection, buoyancy, and dissolution effects play a role in the time period during the injection and the years and decades after the injection period. In order to model this system numerically, a non-isothermal two-phase two-component model including the phases gas and brine as well as the components CO_2 and water needs to be applied. For this system, it is of utmost importance to capture the physical processes occurring correctly in order to make predictions of the fate of the injected CO_2. This especially comprises hysteresis as well as kinetic interphase mass and energy transfer. But also, the considered domain is naturally very large and the physical processes are very complex leading to extremely long computing times. In order to find a good compromise between physical accuracy and computational effort, a multi-physics strategy is proposed to decide where and when an interfacial-area-based model is needed to describe hysteresis, chemical, or thermal non-equilibrium. In the remaining part of the domain, the cheaper classical model may be used. In the following, the characteristic parameters applied to decide about the best model are discussed separately for hysteresis, kinetic interphase mass transfer, and kinetic interphase energy transfer.

Hysteresis Hysteresis may occur mainly due to two processes related to CO_2 injection. First, during the injection period, the porous medium undergoes primary drainage (wetting-phase saturation is decreasing starting from an initially fully brine-saturated porous medium), see left hand side of Fig. 6.4. For that process, the solution of the classical model is sufficient. But then, after the injection is

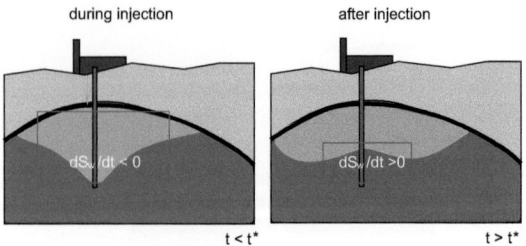

Figure 6.4: Hysteresis due to end of injection.

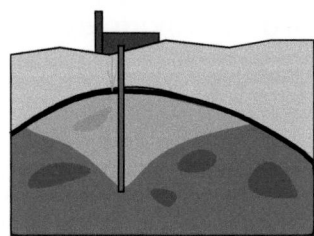

Figure 6.5: Hysteresis due to heterogeneities.

stopped, previously drained regions may be imbibed again, see right hand side of Fig. 6.4. Thus, the sign of $\frac{\partial S_w}{\partial t}$ changes and hysteresis occurs. If that is the case, the interfacial-area-based model needs to be solved. In practice, the sign of $\frac{\partial S_w}{\partial t}$ is to be determined in each element. In the beginning, this sign will be negative throughout the domain. If the sign changes in an element the interfacial-area-based model needs to be solved there.

The second possible reason for hysteresis is the heterogeneity of the porous medium. In reality, the porous medium will not be homogeneous as was suggested by Fig. 6.3. Instead, the subsurface will consist of different materials with different properties, such as different porosity, intrinsic permeability and different entry pressures, see Fig. 6.5. In heterogeneous systems, however, processes such as pooling at fine-material lenses and trapping within coarse-material lenses may occur leading to local changes from drainage to imbibition and vice versa, see Fig. 6.6. In a multi-physics approach, the sign of $\frac{\partial S_w}{\partial t}$ will need to be identified for each element at each time step. If there is a change of the initially negative sign of $\frac{\partial S_w}{\partial t}$ the interfacial-area-based model needs to be solved.

Kinetic interphase mass transfer As the classical model is only able to account for chemical equilibrium situations, a solution of the interfacial-area-based model is necessary if deviations from equilibrium occur. The dimensionless number which

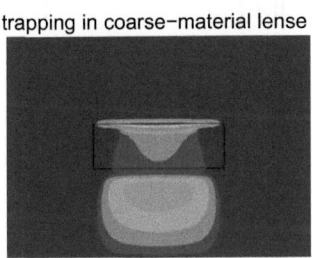

Figure 6.6: Hysteresis due to heterogeneities: pooling and trapping. Here, gas enters an initially fully water-saturated domain from the bottom.

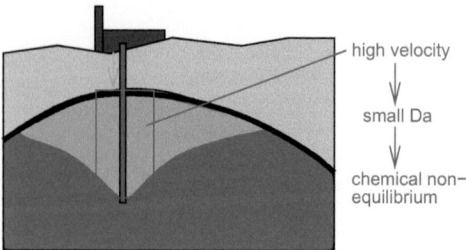

Figure 6.7: Chemical non-equilibrium.

indicates if chemical equilibrium is relevant or not is the Damköhler number that was introduced in Eq. 6.6: For small Damköhler numbers, the interfacial-area-based model needs to be solved. Considering the example of CO_2 storage, high flow velocities will be encountered in the surroundings of the injection well. This in consequence leads to small Damköhler numbers (chemical non-equilibrium) necessitating the solution of the interfacial-area-based model, see Fig. 6.7. In practice, the Damköhler number needs to be calculated at each time step and for each element. If the Damköhler number is smaller than a critical Damköhler number Da_{crit} the interfacial-area-based model needs to be solved; for all $Da > Da_{crit}$, the solution of the classical model is sufficient.

Kinetic interphase energy transfer The classical model is restricted to situations of local thermal equilibrium. In CO_2 injection, however, the injected CO_2 may have a different temperature from the resident rock and brine. This may give rise to the situation of thermal non-equilibrium which can only be captured by use of the interfacial-area-based model. If thermal non-equilibrium is actually relevant or if temperature differences between the phases will equilibrate quickly can be decided by considering the Nusselt number that was defined in Eq. 6.7. If the Nusselt number is small, thermal equilibrium is a good assumption. By considering Fig. 6.8, it becomes obvious that around the injection well flow velocities are high leading to high Nusselt numbers and thermal non-equilibrium. Nusselt numbers larger than a critical Nusselt number Nu_{crit} call for a solution of the interfacial-area-based approach while for all smaller Nusselt numbers the classical model may be solved.

In practice, all of these characteristic parameters and dimensionless numbers need to be evaluated for each element at each time step. If at least one of them calls for a solution of the interfacial-area-based model in a certain subdomain, the interfacial-area-based model is to be applied in this subdomain. With respect to chemical and thermal equilibrium, the switch to the interfacial-area-based model is reversible meaning that conditions may change again such that the classical model can be solved. But the situation is different for hysteresis. If the characteristic parame-

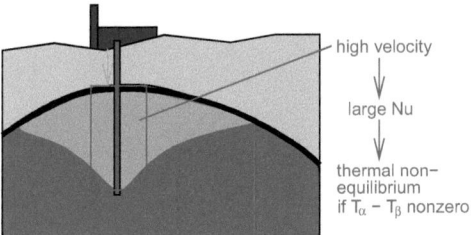

Figure 6.8: Thermal non-equilibrium.

ter for hysteresis (the sign of $\frac{\partial S_w}{\partial t}$) calls for a change in the process, a switch back to the classical model has to be forbidden. The resulting multi-physics algorithm is shown schematically in Fig. 6.9.

6.4 Summary

This chapter focussed on multi-scale multi-physics methods and their advantages with respect to interfacial-area-based modeling. Therefore, current multi-scale, multi-physics, and multi-scale multi-physics approaches were reviewed. Then, based on dimensional analysis and characteristic numbers, it was discussed in detail how to decide when the more expensive interfacial-area-based model should be used and when the less physically based, but cheaper classical model is sufficient. The criteria were discussed in detail with respect to capillary hysteresis, kinetic interphase mass, as well as for kinetic interphase mass and energy transfer. These criteria and the general multi-physics procedure were illustrated using the example of carbon dioxide storage.

```
Set Da_crit ;
Set Nu_crit ;
for (each timestep)
{
    for (each element)
    {
        model = ClassicalModel;
        if (flag == AlwaysAlternativeModel)
        {
            model = AlternativeModel;
            break;
        }
        Calculate SignOfDSwDt;
        if (≠ SignofDSwDtOld)
        {
            model = AlternativeModel;
            SetFlagForElement(AlwaysAlternativeModel);
        }
        Calculate Da;
        if (Da < Da_crit )
            model = AlternativeModel;
        Calculate Nu;
        if (Nu > Nu_crit )
            model = AlternativeModel;
    }
}
```

Figure 6.9: Schematic of the multi-scale algorithm for the example of CO_2 storage.

7 Summary and conclusions

In this chapter, this book is summed up and future research fields evolving from the presented studies are discussed.

7.1 Summary

The main purpose of this thesis is to advance physically based modeling of multi-phase flow and transport processes in porous media by 1) using a thermodynamically consistent set of equations and 2) accounting for the essential role of interfaces in multi-phase flow.

After a general introduction, the classical approach for two-phase flow in porous media including relevant concepts and parameters, balance equations, constitutive relationships, and equations of state was revisited. Several problematic issues of the classical approach were pointed out. Based on the knowledge that phase-interfaces are crucial parameters in two-phase flow in porous media, especially for kinetic mass and heat transfer between phases, and in order to overcome the problems of the classical approach, a need for a thermodynamically consistent approach which explicitly accounts for interfacial areas was derived.

Following an overview of different thermodynamically consistent approaches for two-phase flow, the approach of Hassanizadeh and Gray [1990, 1993a] was chosen as thermodynamic framework as it includes balance equations for not only bulk phases, but also for phase-interfaces and common lines. In the frame of this theory, the averaging of balance equations from the pore to the macro scale was presented and it was shown how the entropy inequality is exploited in order to obtain constitutive relationships. Hassanizadeh and Gray found out that the important classical constitutive relationships for capillary pressure and relative permeability must depend on more parameters than is classically assumed: capillary pressure is not a function of saturation only (as is classically assumed), but additionally depends on the interfacial areas. Relative permeability is also not a function of saturation only; instead, the thermodynamically consistent approach suggests that the effect of missing driving forces (saturation gradients) in the classical extended Darcy's law are lumped into relative permeability.

Based on the thermodynamically consistent set of equations including interfacial area, mathematical models of different complexity were developed. In this development process, partly, the thermodynamically consistent set of equations needed to be simplified, and partly, it needed to be supplemented by new balance equations for components and / or energy which themselves needed to be upscaled. In order to solve the resulting set of governing equations, a numerical method needs to be selected. As it is locally mass-conservative and extendible to unstructured grids, the vertex-centered finite volume method was chosen for the simulations of the interfacial-area-based approach in this work. Specifically, the following numerical models were developed and tested:

- **Isothermal immiscible incompressible two-phase flow—without hysteresis**
  For that case, the classical two-phase model and the interfacial-area-based model should give exactly the same results. Comparing the results of the two models for a primary drainage process in a horizontal porous medium, it was verified that both models are in excellent agreement (Niessner and Hassanizadeh [2008]).

- **Isothermal immiscible incompressible two-phase flow—with hysteresis**
  In the next step, an interfacial-area-based semi-analytical solution for Philip's problem (Philip [1991]) was sought. Such a semi-analytical solution that allows for a verification of the numerical model, was derived (Pop et al. [2009]). In a Dipl.-Ing. thesis (Marshall [2009]), a coupling algorithm was developed which allows to solve Philip's problem numerically using the interfacial-area-based approach. A very good agreement between the numerical model and the semi-analytical solution was found.

- **Isothermal two-phase flow with kinetic interphase mass transfer**
  Kinetic interphase mass transfer can only be described in a physically based way if the specific fluid–fluid interfacial area is known. This is due to the fact that kinetic interphase mass transfer is proportional to the deviation from local chemical equilibrium and restricted by the available interfacial area. The set of governing equations was developed and implemented into a numerical simulator (Niessner and Hassanizadeh [2009a]). Simulation results show a significant effect of chemical non-equilibrium in case of injection of dry air into a highly water-saturated porous medium. This effect could only be captured by the interfacial-area-based model, but not by the classical model.

- **Two-phase flow with kinetic interphase mass and heat transfer**
  Kinetic interphase heat transfer can only be described in a physically based way if the specific fluid–fluid interfacial area, but also the specific solid-fluid interfacial areas are known. This is due to the fact that unlike mass, heat may additionally be transferred from and to the solid phase. Pore-scale energy balance equations were upscaled to the macro-scale and the complete set of governing equations was implemented into a numerical simulator. Example simulations of the processes in an evaporator and of drying of a porous medium by injection of hot dry air showed that the effect of thermal non-equilibrium may be significant (Niessner and Hassanizadeh [2009c] and Ahrenholz et al. [2009]). This effect, however, cannot be captured in a physically based way

unless interfacial areas are explictly included as parameters.

- **Continuity conditions at material interfaces**
  The physically correct treatment of macro-scale material interfaces represents a major challenge: while variables, such as pressures, specific interfacial area, and fluxes are continuous across such material interfaces, saturation is generally discontinuous. The representation of discontinuities, however, is numerically extremely difficult. Therefore, an interface condition was proposed that allows for the correct treatment of such material interfaces.

- **Gas–water flow in a fracture–matrix system**
  Fractured porous media are characterized by two facts: on the one hand, flow in fractures is fast, generally leading to local chemical non-equilibrium; on the other hand, a fractured porous medium is intrinsically heterogeneous which always leads to hysteresis. Both aspects can only be correctly described if interfacial areas are included. Therefore, gas–water flow in a fracture–matrix system was chosen as a real-life application where interfacial areas are extremely relevant. In his Dipl.-Ing. thesis, Nuske [2009] obtained constitutive relationships for fractures from a micro-scale model and Faigle [2009] (another Dipl.-Ing. thesis) performed macro-scale simulations using these relationships. It could be shown that hysteresis and kinetic interphase mass transfer are indeed relevant and can be described numerically through including interfacial area as parameter.

In order to make realistic predictions using the interfacial-area-based approach, new parameters and constitutive relationships need to be determined. Therefore, different experimental and numerical methods are presented in this thesis that allow to obtain relevant parameters and constitutive relationships for the interfacial-area-based model.

As the interfacial-area-based approach is computationally more expensive than the classical approach, a compromise to handle the dilemma between physical consistency and computational efficiency was sought. It was found that multi-scale multi-physics (multi-numerics) approaches represent a technique to find such a compromise. Therefore, as a perspective, it was shown how the interfacial-area-based approach can be nicely fit into such a multi-scale multi-physics framework. In order to decide whether hysteresis occurs necessitating the use of the interfacial-area-based model, the sign of the time rate of change of saturation was used as an indicator. With respect to kinetic interphase mass and mass and energy transfer, the respective equations were non-dimensionalized and dimensionless numbers were identified which help to decide whether the interfacial-area-based model needs to be solved or whether the classical model gives sufficiently good results.

7.2 Outlook

Future research concerning thermodynamically consistent modeling of two-phase flow in porous media accounting for the role of interfacial areas is necessary in many different directions. They comprise basic research on the thermodynamically based theory, experimental investigation on the "correctness" of the thermodynamically consistent theory, determination of parameters and constitutive relationships, the adaptation of the interfacial-area-based approach to more applications where interfacial areas are relevant, and the solution of application-oriented problems. In the following, these issues will be discussed in more detail.

- **Basic research on the thermodynamically consistent theory**
  As the thermodynamically consistent theory is the fundament for all numerical studies and experimental investigations on parameters and constitutive relationships in the frame of the interfacial-area-based approach, it needs to be studied thoroughly. The upscaling (volume averaging) of conservation equations from the pore scale to the macro scale is rigorous. Also, the formulation of the entropy inequality results directly and clearly from the macro-scale balance equations. However, it is not clear a priori on which parameters the Helmholtz free energy of the phases and interfaces depends. In principle, it may depend on the whole long list of primary variables. The assumption on which parameters it primarily depends and which dependencies might be neglected is crucial and determines the final macro-scale balance equations containing the constitutive relationships. Therefore, the influence of assuming Helmholtz free energies of phases to depend on more or less primary variables than saturation, density, and temperature (and Helmholtz free energies of interfaces additionally on specific interfacial areas) is to be investigated. Also, it is very important to compare the procedures and the final macro-scale set of balance equations in the frame of different thermodynamic approaches (e.g. rational thermodynamics, irreversible thermodynamics, thermodynamically constrained averaging) for which the same set of parameters is chosen.

- **Experimental investigations on the "correctness" of the thermodynamically consistent theory**
  In order to strengthen the fundament of the thermodynamically consistent theory and to manifest its abilities in terms of physical correctness experimental verification is inevitable. Experiments need to be designed which allow to determine whether additional terms in the momentum balance occur compared to Darcy's law, and to investigate the relevance of non-classical effects, such as non-equilibrium effects in the difference in phase pressures.

- **Determination of parameters and constitutive relationships**
  As the interfacial-area-based approach contains new parameters and constitutive relationships carefully designed experiments are essential. Parameters to be determined include interfacial permeability, the production rate of in-

terfacial area, and diffusion lengths. Constitutive relationships which need investigating are the relationship among specific interfacial areas (fluid–fluid and fluid–solid), capillary pressure, and saturation. Ideally, a cheap and easy-to-use measurement procedure is to be designed that allows to determine this relationship in a standardized manner.

- **Solution of application-oriented problems**
  Several highly important real-life problems cannot be numerically modeled in a physically based way without accounting for the role of interfacial areas. Among those are carbon dioxide storage in deep geological formations, gas migration in fracture–matrix systems, non-isothermal hysteretic flow and transport processes in the unsaturated zone of the subsurface, methane migration, investigation on the processes in fuel cells, as well as the application of the interfacial-area-based model to industrial porous media such as drying of porous media where thermal non-equilibrium processes may become significant. For all these applications, the interfacial-area-based models presented in this work can be directly used.

- **Adaptation of the interfacial-area-based approach to more applications where interfacial areas are relevant**
  In many disciplines, such as groundwater remediation through use of surfactants, bacteria and virus transport, or relatively fast flow interfacial areas play a major and dominating role. However, the concepts of this work must be extended to allow for a description of these systems. Flow and transport processes in porous media will be identified where interfacial areas allow for a physically based modeling and where to date, only an empirical model could be formulated. These comprise the modeling of surfactants (tensides) in two-phase flow in porous media. These surfactants sit on the fluid–fluid interfaces by definition and ideally, form a mono-molecular layer there. Thus, the concentration of surfactants in e.g. the bulk water phase is closely related to the amount of interfacial area that is present. Microorganisms, such as bacteria and viruses show a similar behavior. Another important extension of the thermodynamically consistent model is that for non-equilibrium effects in the difference in phase pressures. This effect has been implemented, but has not been thoroughly investigated numerically.

This book is a contribution to thermodynamically consistent modeling of two-phase flow in porous media accounting for the important role of phase-interfaces. Such a thermodynamically consistent model allows to deepen the understanding of the physical processes involved and allows to make physically based predictions. Instead of extending simple models, a complex, thermodynamically consistent model was used, adapted to special applications, and simulations were performed. With permanently increasing computer power and with measurement of interfacial area being a growing field of research, thermodynamically consistent modeling accounting for the important role of interfacial area will become possible for realistic large-scale applications, such as CO_2 storage in the near future. This will allow for a thorough problem understanding and for physically based predictions.

Bibliography

L.M. Abriola and G.F. Pinder. A multiphase approach to the modeling of porous media contamination by organic compounds, 1, Equation development. *Water Resources Research*, 21(1):11–18, 1985a.

L.M. Abriola and G.F. Pinder. A multiphase approach to the modeling of porous media contamination by organic compounds, 2, Numerical simulation. *Water Resources Research*, 21(1):19–26, 1985b.

B. Ahrenholz, J. Niessner, , R. Helmig, and M. Krafczyk. Determination of parameters for modeling evaporation processes in porous media. *Advances in Water Resources*, 2009. submitted.

M. Allen. Numerical modeling of multiphase flow in porous media. *Advances in Water Resources*, 8:162–187, 1985.

W.F. Ames. *Numerical Methods for Partial Differential Equations*. Academic Press, New York, 1977.

F.V. Atkinson and L.A. Peletier. Similarity profiles of flows through porous media. *Archive for Rational Mechanics and Analysis*, 42:369–379, 1971.

F.V. Atkinson and L.A. Peletier. Similarity solutions of the nonlinear diffusion equation. *Archive for Rational Mechanics and Analysis*, 54:373–392, 1974.

J.L. Auriault. Transport in porous media : upscaling by multiscale asymptotic expansions. In *Applied Micromechanics of Porous Materials*, pages 3–56. Springer, 2005.

D.G. Avraam and A.C. Payatakes. Generalized relative permeability coefficients during steady-state two-phase flow in porous media, and correlation with the flow mechanisms. *Transport in Porous Media*, 20(1–2):135–168, 1995a.

D.G. Avraam and A.C. Payatakes. Flow regimes and relative permeabilities during steady-state two-phase flow in porous media. *Journal of Fluid Mechanics Digital Archive*, 293:207–236, 1995b.

K. Aziz and A. Settari. *Petroleum Reservoir Simulation*. Elsevier Applied Science, 1979.

A. Baehr and M.Y. Corapcioglu. A predictive model for pollution from gasoline in Soils and Groundwater. In *Proceedings of the NWWA/API Conference in Petroleum Hydrocarbons and Organic Chemicals in Ground WaterJournal of Engineering Mechanics*, pages 144–156, 1984.

P. Bastian. Numerical Computation of Multiphase Flows in Porous Media. Habilitationsschrift vorgelegt an der Technischen Fakultät der Christian–Albrechts–Universität Kiel, 1999.

P. Bastian and R. Helmig. Efficient Fully-Coupled Solution Techniques for Two Phase Flow in Porous Media. Parallel Multigrid Solution and Large Scale Computations. *Advances in Water Resources*, 23:199–216, 1999.

P. Bastian, K. Birken, K. Johannsen, S. Lang, K. Eckstein, N. Neuss, H. Rentz-Reichert, and C. Wieners. UG - A Flexible Software Toolbox for Solving Partial Differential Equations. Computing and Visualization in Science, 1(1):27–40, 1997.

J. Bear. *Dynamics of Fluids in Porous Media*. Elsevier, New York, 1972.

J. Bear and Y. Bachmat. *Introduction to Modeling of Transport Phenomena in Porous Media*. Kluwer Academic Publishers, The Netherlands, 1990.

S. Berg, A.W. Cense, J.P. Hofman, and R.M.M. Smits. Two-Phase Flow in Porous Media with Slip Boundary Condition. *Transport in Porous Media*, 74:275–292, 2008.

P. Binning and M. A. Celia. Practical implementation of the fractional flow approach to multi-phase flow simulation. *Advances in Water Resources*, 22(5):461–478, 1999.

P. Binning and M.A. Celia. A finite volume eulerian-lagrangian localized adjoint method for solution of the contaminant transport equations in two-dimensional multi-phase flow systems. *Water Resources Research*, 32:103–114, 1996.

S. Bottero, S.M. Hassanizadeh, P.J. Kleingeld, and A. Bezuijen. Experimental study of dynamic capillary pressure effect in two-phase flow in porous media. In *Proceedings of the XVI International Conference on Computational Methods in Water Resources (CMWR), Copenhagen, Denmark*, 2006.

R.M. Bowen. Compressible Porous Media Models by Use of the Theory of Mixtures. *International Journal of Engineering Science*, 20(6):697–735, 1982.

H.C. Brinkman. A calculation of the viscous force exerted by a flowing fluid on a dense swarm of particles. *Applied Science Research*, A1:27–33, 1947.

A. N. Brooks and A. T. Corey. *Hydraulic Properties of Porous Media*. Colorado State University, 1964.

M. Brusseau, J. Popovicova, and J. Silva. Characterizing gas–water interfacial and bulk-water partitioning for gas phase transport of organic contaminants in unsaturated porous media. *Environmental Sciences Technology*, 31:1645–1649, 1997.

E. Buckingham. Studies on the movement of soil moisture. Technical Report Bull. 38, Bureau of Soils, Washington, DC, 1907. USDA.

J.S. Buckley. Multiphase displacements in micromodels. In *Interfacial Phenomena in Petroleum Recovery*, pages 157–189. Marcel Dekker, New York, 1991.

N.T. Burdine. Relative permeability calculation from size distribution data. *Transactions of the AIME*, 198:71–78, 1953.

G. Chavent and J. Jaffré. *Mathematical models and finite elements for reservoir simulation*. North Holland, Amsterdam, 1986.

D. Chen, L.J. Pyrak-Nolte, J. Griffin, and N. Giordano. Measurement of interfacial area per volume for drainage and imbibition. *Water Resources Research*, 43(12), 2007.

L. Chen and T. Kibbey. Measurement of air–water interfacial area for multiple hysteretic drainage curves in an unsaturated fine sand. *Langmuir*, 22:6674–6880, 2006.

Y. Chen, L. J. Durlofsky, M. Gerritsen, and X.H. Wen. A coupled local-global upscaling approach for simulating flow in highly heterogeneous formations. *Advances in Water Resources*, 26:1041–1060, 2003.

Z. Chen and R. E. Ewing. Fully Discrete Finite Element Analysis of Multiphase Flow in Groundwater Hydrology. *SIAM Journal of Numerical Analysis*, 34(6):2228–2253, 1997.

Z. Chen and T. Y. Hou. A mixed multiscale finite element method for elliptic problems with oscillating coefficients. *Mathematics of Computation*, 72(242):541–576, 2002.

K. Coats. An equation of state compositional model. *Society of Petroleum Engineering Journal*, 20:363–376, 1980.

D. Crandall, G. Ahmadi, D. Leonard, M. Ferer, and D.H. Smith. A new stereolithography experimental porous flow device. *Review of Scientific Instruments*, 79(044501), 2008.

S. Crone, C. Bergins, and K. Strauss. Multiphase Flow in Homogeneous Porous Media with Phase Change. Part I: Numerical Modeling. *Transport in Porous Media*, 49:291–312, 2002.

K. Culligan, D. Wildenschild, B. Christensen, W. Gray, M. Rivers, and A. Tompson. Interfacial area measurements for unsaturated flow through a porous medium. *Water Resources Research*, 40:1–12, 2004.

K.A. Culligan, D. Wildenschild, B.S.B Christensen, W.G. Gray, and M.L. Rivers. Pore-scale characteristics of multiphase flow in porous media: a comparison of air–water and oil–water experiments. *Advances in Water Resources*, 29:227–38, 2006.

H. Darcy. Détermination des lois d'écoulement de l'eau à travers le sable. In *Les Fontaines Publiques de la Ville de Dijon*, pages 590–594. Victor Dalmont, Paris, 1856.

A.H. Demond and P.V. Roberts. An examination of relative pereability relations for two-phase flow in porous media. *Journal of the American Water Resources Association*, 23(4):617–628, 1987.

D.A. DiCarlo. Experimental measurements of saturation overshoot on infiltration. *Water Resources Research*, 40(W04215), 2004.

D.A. DiCarlo. Modeling observed saturation overshoot with continuum additions to standard unsaturated theory. *Advances in Water Resources*, 28(10):1021–1027, 2005.

L. J. Durlofsky. Numerical Calculation of Equivalent Grid Block Permeability Tensors for Heterogeneous Porous Media. *Water Resources Research*, 27(5):699–708, 1991.

F. Duval, F. Fichot, and M. Quintard. A local thermal non-equilibrium model for two-phase flows with phase-change in porous media. *International Journal of Heat and Mass Transfer*, 47:613–639, 2004.

W. E and B. Engquist. Multiscale modeling and computation. *Notices of the AMS*, 50(9):1062–1070, 2003a.

W E. and B. Engquist. The heterogeneous multiscale methods. *Commun. Math. Sci.*, 1(1):87–132, 2003b. ISSN 1539-6746.

W. E, B. Engquist, and Z. Huang. Heterogeneous multiscale method: A general methodology for multiscale modeling. *Physical Review*, 67, 2003.

W. E, B. Engquist, X. Li, W. Ren, and E. Vanden-Eijnden. Heterogeneous multiscale methods: a review. *Commun. Comput. Phys.*, 2(3):367–450, 2007. ISSN 1815-2406.

Y. Efendiev and L. J. Durlofsky. Numerical modeling of subgrid heterogeneity in two phase flow simulations. *Water Resources Research*, 38(8), 2002.

Y. Efendiev, L. J. Durlofsky, and S. H. Lee. Modeling of subgrid effects in coarse-scale simulations of transport in heterogeneous porous media. *Water Resources Research*, 36(8):2031 – 2041, 2000.

R.E. Ewing and H. Wang. Eulerian-lagrangian localized adjoint methods for variable-coefficient advective-diffusive-reactive equations in groundwater contaminant transport. In *Gomez, Hennart (Eds.), Advances in Optimization and Numerical Analysis, Mathematics and Its Applications*, pages 185–205. Kluwer Academic Publishers, Dordrecht, Netherlands, 1994.

B. Faigle. Two-phase flow modeling in porous media with kinetic interphase mass transfer processes in fractures. Dipl.-Ing. thesis, Institute of Hydraulic Engineering, University of Stuttgart, 2009.

R. W. Falta. Numerical modeling of kinetic interphase mass transfer during air sparging using a dual-media approach. *Water Resources Research*, 36(12):3391–3400, 2000.

R. W. Falta. Modeling sub-grid-block-scale dense nonaqueous phase liquid (DNAPL) pool dissolution using a dual-domain approach. *Water Resources Research*, 39(12), 2003.

J. Fritz, B. Flemisch, and R. Helmig. Multiphysics modeling of advection-dominated two-phase compositional flow in porous media. *International Journal of Numerical Analysis and Modeling*, 2009. accepted.

R. Fucik, Mikyska J., and M. Illangasekare T.H. Benes. Semi-analytical solution for two-phase flow in porous media with a discontinuity. *Vadose Zone Journal*, 7:100–1007, 2008.

G.P. Grant and J.I. Gerhard. Simulating the dissolution of a complex DNAPL source zone: 2. Experminental validation of an interfacial area-based mass transfer model. *Water Resources Research*, 43(W12409):1–18, 2007. doi:10.1029/2007WR006039.

W. Gray and S. Hassanizadeh. Averaging theorems and averaged equations for transport of interface properties in multiphase systems. *International Journal of Multi-Phase Flow*, 15:81–95, 1989.

W.G. Gray and S.M. Hassanizadeh. Paradoxes and realities in unsaturated flow theory. *Water Resources Research*, 27(8):1847–1854, 1991.

W.G. Gray and S.M. Hassanizadeh. Macroscale continuum mechanics for multiphase porous-media flow including phases, interfaces, common lines, and common points. *Advances in Water Resources*, 21:261–281, 1998.

W.G. Gray and C.T. Miller. Thermodynamically Constrained Averaging Theory Approach for Modeling of Flow in Porous Media: 1. Motivation and Overview. *Advances in Water Resources*, 28(2):161–180, 2005.

S. M. Hassanizadeh and W. G. Gray. General Conservation Equations for Multi-Phase Systems: 1. Averaging Procedure. *Advances in Water Resources*, 2:131–144, 1979a.

S. M. Hassanizadeh and W. G. Gray. General Conservation Equations for Multi-Phase Systems: 2. Mass, Momenta, Energy, and Entropy Equations. *Advances in Water Resources*, 2:191–203, 1979b.

S. M. Hassanizadeh and W. G. Gray. General Conservation Equations for Multi-Phase Systems: 3. Constitutive Theory for Porous Media Flow. *Advances in Water Resources*, 3:25–40, 1980.

S. M. Hassanizadeh and W. G. Gray. Mechanics and Thermodynamics of Multiphase Flow in Porous Media Including Interphase Boundaries. *Advances in Water Resources*, 13(4):169–186, 1990.

S. M. Hassanizadeh and W. G. Gray. Toward an improved description of the physics of two-phase flow. *Advances in Water Resources*, 16(1):53–67, 1993a.

S. M. Hassanizadeh and W. G. Gray. Thermodynamic Basis of Capillary Pressure in Porous Media. *Water Resources Research*, 29(10):3389 – 3405, 1993b.

R. Held and M. Celia. Modeling support of functional relationships between capillary pressure, saturation, interfacial area and common lines. *Advances in Water Resources*, 24:325–343, 2001.

R. Helmig. *Multiphase Flow and Transport Processes in the Subsurface*. Springer, 1997.

R. Helmig, J. Niessner, B. Flemisch, M. Wolff, and J. Fritz. Efficient modelling of flow and transport in porous media using multi-physics and multi-scale approaches. In *Handbook of Geomathematics*. Springer, 2010.

R. Hilfer. Macroscopic Capillarity and Hysteresis for Flow in PorousMedia. *Physical Review E*, 73(016307), 2006.

C. Hirsch. *Numerical Computation of Internal and External Flows; Volume 1: Fundamentals of Numerical Discretization*. John Wiley & Sons, New York, 1988.

T. Hou and X. H. Wu. A multiscale finite element method for elliptic problems in composite materials and porous media. *Journal of Computational Physics*, 134:169–67, 1997.

R. U. Huber. *Compositional Multiphase Flow and Transport in Heterogeneous Porous Media*. PhD thesis, Report no. 102, Institute of Hydraulic Engineering, University of Stuttgart, 2000. ISBN: 3-933761-05-0.

IAPWS (The International Association for the Properties of Water and Steam). Revised Release on the IAPS Formulation 1985 for the Viscosity of Ordinary Water Substance, 2003. http://www.iapws.org.

P.T. Imhoff, P.R. Jaffe, and G.F. Pinder. An experimental study of complete dissolution of a nonaqueous phase liquid in saturated porous media. *Water Resources Research*, 30(2):307–320, 1994.

IPCC. *Carbon Dioxide Capture and Storage. Special Report of the Intergovernmental Panel on Climate Change*. Cambridge University Press, 2005.

H. Jakobs. *Simulation nicht-isothermer Gas–Wasser-Prozesse in komplexen Kluft–Matrix-Systemen*. PhD thesis, Institute of Hydraulic Engineering, University of Stuttgart, 2004. ISBN: 3-922761-31-X.

P. Jenny, S. H. Lee, and H. Tchelepi. Multi-scale finite-volume method for elliptic problems in subsurface flow simulations. *Journal of Computational Physics*, 187: 47–67, 2003.

G.R. Jerauld and S.J. Salter. The effect of pore-structure on hysteresis in relative permeability and capillary pressure: Pore-level modeling. *Transport in Porous Media*, 5(2):103–151, 1990.

V. Joekar-Niasar, S. M. Hassanizadeh, and A. Leijnse. Insights into the relationship among capillary pressure, saturation, interfacial area and relative permeability using pore-scale network modeling. *Transport in Porous Media*, 74:201–219, 2008.

V. Joekar-Niasar, S. M. Hassanizadeh, L. J. Pyrak-Nolte, and C. Berentsen. Simulating drainage and imbibition experiments in a high-porosity micro-model using an unstructured pore-network model. *Water Resources Research*, 45(W02430, doi:10.1029/2007WR006641), 2009.

F. Kalaydjian. A Macroscopic Description of Multiphase Flow in Porous Media Involving Spacetime Evolution of Fluid/Fluid Interface. *Transport in Porous Media*, 2:537 – 552, 1987.

S. E. Koch. A Survey of Nested Grid Techniques and Their Potential for Use Within the MASS Weather Prediction Model. *NASA Technical Memorandum 87808*, 1987.

S. Korteland, S. Bottero, S.M. Hassanizadeh, and C.W.J. Berentsen. What is the correct definition of macroscale pressure? *Transport in Porous Media*, 2009. accepted for publication.

L. Lake. *Enhanced Oil Recovery*. Prentice-Hall, Inc., 1989.

M.G. Larson and A. Målqvist. An adaptive variational multiscale method for convection-diffusion problems. *Comm. Numer. Methods Engrg.*, 25(1):65–79, 2009. ISSN 1069-8299.

E.J. Lefebvre du Prey. Factors affecting liquid-liquid relative permeabilities of a consolidated porous medium. *Society of Petroleum Engineering Journal*, 13(1):39–47, 1973.

R. Lenormand, C. Zarcone, and A. Sarr. Mechanisms of the displacement of one fluid by another in a network of capillary ducts. *Journal of Fluid Mechanics*, 135: 337–353, 1983.

M.C. Leverett. Flow of Oil-Water Mixtures through Unconsolidated Sands. AIME Petroleum Transactions, Band 142, S. 152-169, 1941.

B.B. Looney and R.W. Falta. *Vadose Zone*. Battelle Press, Columbus, OH, 2000.

C. Lüdecke and D. Lüdecke. *Thermodynamik*. Springer, Berlin, 2000.

S. Ma, G. Mason, and N.R. Morrow. Effect of Contact Angle on Drainage and Imbibition in Regular Polygonal Tubes. *Colloids and Surfaces A*, 117:273–291, 1996.

S. Manthey, M. S. Hassanizadeh, and R. Helmig. Macro-scale dynamic effects in homogeneous and heterogeneous porous media. *Transport in Porous Media*, 58: 121–145, 2005.

C.-M. Marle. From the pore scale to the macroscopic scale: Equations governing multiphase fluid flow through porous media. In *Proceedings of Euromech 143*, pages 57–61, 1981. Delft, Verruijt, A. and Barends, F. B. J. (eds.).

F. Marshall. Numerical Solution of Philip's Redistribution Problem. Dipl.-Ing. thesis, Institute of Hydraulic Engineering, University of Stuttgart, 2009.

A.S. Mayer and S.M. Hassanizadeh. *Soil and Groundwater Contamination: Nonaqueous Phase Liquids*. American Geophysical Union, 2005.

R.P. Mayer and R.A. Stowe. Mercury Porosimetry—Breakthrough Pressure for Penetration Between Packed Spheres. *Journal of Colloid Science*, 20:893–911, 1965.

C. T. Miller, G. Christakos, P. T. Imhoff, J. F. McBride, J. A. Pedit, and J. A. Trangenstein. Multiphase flow and transport modeling in heterogeneous porous media: challenges and approaches. *Advances in Water Resources*, 21(2):77–120, 1998.

C.T. Miller and W.G. Gray. Thermodynamically constrained averaging theory approach for modeling flow and transport phenomena in porous medium systems: 2. foundation. *Advances in Water Resources*, 28(2):181–202, 2005.

C.T. Miller, M.M. Poirier-McNeill, and A.S. Mayer. Dissolution of trapped nonaqueous phase liquids: Mass transfer characteristics. *Water Resources Research*, 21(2): 77–120, 1990.

R.J. Millington and J.P. Quirk. Permeability of Porous Solids. *Transactions of the Faraday Society*, 57:1200–1207, 1961.

M. Mirzaei and D. B. Das. Dynamic effects in capillary pressure–saturations relationships for two-phase flow in 3d porous media: Implications of micro-heterogeneities. *Chemical Engineering Science*, 62(7):1927–1947, 2007.

Y. Mualem. A new model for predicting the hydraulic conductivity of unsaturated porous media. *Water Resources Research*, 12:513–522, 1976.

J. Niessner and S.M. Hassanizadeh. A Model for Two-Phase Flow in Porous Media Including Fluid–Fluid Interfacial Area. *Water Resources Research*, 2008. 44, W08439, doi:10.1029/2007WR006721.

J. Niessner and S.M. Hassanizadeh. Modeling kinetic interphase mass transfer for two-phase flow in porous media including fluid–fluid interfacial area. *Transport in Porous Media*, 2009a. doi:10.1007/s11242-009-9358-5.

J. Niessner and S.M. Hassanizadeh. Two-phase flow and transport in porous media including fluid–fluid interfacial area. In S. Martin and J.R. Williams, editors, *Multiphase Flow Research*. Novascience Publisher, 2009b.

J. Niessner and S.M. Hassanizadeh. Non-equilibrium interphase heat and mass transfer during two-phase flow in porous media—theoretical considerations and modeling. *Advances in Water Resources*, 32:1756–1766, 2009c.

J. Niessner and R. Helmig. Multi-scale modeling of two-phase - two-component processes in heterogeneous porous media. *Numerical Linear Algebra with Applications*, 13(9):699–715, 2006.

J. Niessner and R. Helmig. Multi-scale modeling of three-phase–three-component processes in heterogeneous porous media. *Advances in Water Resources*, 11(30): 2309–2325, 2007.

J. Niessner and R. Helmig. Multi-physics modeling of flow and transport in porous media using a downscaling approach. *Advances in Water Resources*, 32(6):845–850, 2009.

J. Niessner, S.M. Hassanizadeh, and D. Crandall. Modeling two-phase flow in porous media including fluid–fluid interfacial area. In E. Bänsch, editor, *Proceedings of the ASME IMECE congress*, Boston, Massachusetts, 2008.

J. M. Nordbotten and P. E. Bjørstad. On the relationship between the multiscale finite-volume method and domain decomposition preconditioners. *Comput. Geosci.*, 12(3):367–376, 2008. ISSN 1420-0597.

J.M. Nordbotten, M.A. Celia, H.K. Dahle, and S.M. Hassanizadeh. On the definition of macro-scale pressure for multi-phase flow in porous media. *Water Resources Research*, 44(6, W06S02, doi:10.1029/2006WR005715), 2008.

H. Nordhaug, M. Celia, and H. Dahle. A pore network model for calculation of interfacial velocities. *Advances in Water Resources*, 26:1061–1074, 2003a.

H. Nordhaug, H. Dahle, M. Espedal, W. Gray, and M. Celia. Two-phase flow including interfacial area as a variable. In L. Bentley, C. Brebbia, W. Gray, and G. Pinder, editors, *Computational Methods in Water Resources*, volume 1, pages 231–238, 2003b.

P. Nuske. Determination of Interfacial Area–Capillary Pressure–Saturation Relationships for a Single Fracture. Dipl.-Ing. thesis, Institute of Hydraulic Engineering, University of Stuttgart, 2009.

S. Ochs, H. Class, R. Helmig, and M. Acosta. Simulation of multiphase multicomponent processes in the diffusion layer of PEM fuel cells. In M. Qintard, editor, *Proceedings of Eurotherm 81*, Albi, France, 2007.

J. C. Parker, R. J. Lenhard, and T. Kuppusami. A Parametric Model for Constitutive Properties Governing Multiphase Flow in Porous Media. *Water Resources Research*, 23(4):618–624, 1987.

D. W. Peaceman. *Fundamentals of Numerical Reservoir Engineering*. Elsevier Applied Science, 1977.

M. Peszynska, M. F. Wheeler, and I. Yotov. Mortar upscaling for multiphase flow in porous media. *Computational Geosciences*, 6:73–100, 2002.

J. R. Philip. Horizontal redistribution with capillary hysteresis. *Water Resources Research*, 27:1459–1469, 1991.

D. W. Pollock. Semi-Analytical Computation of Path Lines for Finite Difference Models. *Ground Water*, 26(6):743–750, 1988.

I.S. Pop, C.J. van Duijn, J. Niessner, and S.M. Hassanizadeh. Horizontal redistribution of fluids in a porous medium: The role of interfacial area in modeling hysteresis. *Advances in Water Resources*, 32(3):383–390, 2009.

M.L. Porter, M.G. Schaap, and D. Wildenschild. Lattice-boltzmann simulations of the capillary pressure-saturation-interfacial area relationship for porous media. *Advances in Water Resources*, 32(11):1632–1640, 2009.

S.E. Powers, L.M. Abriola, and W.J. Weber. An experimental investigation of non-aqueous phase liquid dissolution in saturated subsurface systems: steady state mass transfer rates. *Water Resources Research*, 28:2691–2706, 1992.

S.E. Powers, L.M. Abriola, and W.J. Weber. An experimental investigation of non-aqueous phase liquid dissolution in saturated subsurface systems: transient mass transfer rates. *Water Resources Research*, 30(2):321–332, 1994.

J. M. Prausnitz, R. N. Lichtenthaler, and E. G. Azevedo. *Molecular Thermodynamics of Fluid-Phase Equilibria*. Prentice-Hall, 1967.

H. M. J. Princen. Capillary Phenomena in Assemblies of Parallel Cylinders I. Capillary Rise between Two Cylinders. *Colloid Interface Science*, 30, 1969.

L. J. Pyrak-Nolte, D. D. Nolte, D. Chen, and N.J. Giordano. Relating Capillary Pressure to Interfacial Areas. *Water Resources Research*, 44(W06408), 2008. doi:10.1029/2007WR006434.

P. Reeves and M. Celia. A functional relationship between capillary pressure, saturation, and interfacial area as revealed by a pore-scale network model. *Water Resources Research*, 32(8):2345–2358, 1996.

V. Reichenberger, H. Jakobs, P. Bastian, R. Helmig, and J. Niessner. Complex Gas-Water Processes in Discrete Fracture-Matrix Systems. Upscaling, Mass-Conservative Discretization and Efficient Multilevel Solution and Efficient Multilevel Solution. Mitteilungsheft, Institut für Wasserbau, Universität Stuttgart, 2004. 3-9337 61-33-6.

P. Renard and G. de Marsily. Calculating effective permeability: a review. *Advances in Water Resources*, 20:253 – 278, 1997.

L.A. Richards. Capillary conduction of liquids through porous mediums. *Physics*, 1: 318–333, 1931.

W. Rose. Myths about later-day extensions of Darcy's law. *Journal of Petroleum Sciences Engineering*, 26:187–198, 2000.

T. F. Russell. Time stepping along characteristics with incomplete ite- ration for a galerkin approximation of miscible displacement in porous media. *SIAM Journal of Numerical Analysis*, 22:970–1013, 1985.

C. Schaefer, D. DiCarlo, and M. Blunt. Experimental measurements of air–water interfacial area during gravity drainage and secondary imbibition in porous media. *Water Resources Research*, 36:885–890, 2000.

A.E. Scheidegger. *The Physics of Flow through Porous Media*. University of Toronto Press, Toronto, 3. edition, 1974.

D.K. Schwartz, M.L. Schlossman, E.H. Kawamoto, G.J. Kellogg, P.S. Pershan, and B.M. Ocko. Thermal diffuse x-ray scattering studies of the water–vapor interface. *Physical Review Letters A*, 41(10):5687–5690, 1990.

J.H. Smith and J.A.C. Humphrey. Modeling convection enhanced delivery in brain: Fluid and mass transport analysis. In *9th AIAA/ASME Joint Thermophysics and Heat Transfer Conference, 5-8 June 2006, San Francisco, California*, 2006.

F. Stauffer. Time dependence of the relations between capillary pressure, water content and conductivity during drainage of porous media. In *On scale effects in porous media*, Thessaloniki, Greece, 1978. IAHR.

T.K. Tokunaga and J. Wan. Water film flow along fractures of porous rock. *Water Resources Research*, 33(6):1287–1295, 1997.

D.J. van Antwerp, R.W. Falta, and J.S. Gierke. Numerical Simulation of Field-Scale Contaminant Mass Transfer during Air Sparging. *Vadose Zone Journal*, 7:294–304, 2008.

C.J. van Duijn and M.J. de Neef. Similarity solution for capillary redistribution of two phases in a porous medium with a single discontinuity. *Advances in Water Resources*, 21:451–461, 1998.

C.J. van Duijn and L.A. Peletier. A class of similarity solutions of the nonlinear diffusion equation. *Nonlinear Analysis*, 1:223–233, 1976/77.

M. T. van Genuchten. A Closed-Form Equation for Predicting the Hydraulic Conductivity of Unsaturated Soils. *American Journal of Soil Science*, 44:892–898, 1980.

M.F. Wheeler, T. Arbogast, S. Bryant, J. Eaton, Q. Lu, M. Peszynska, and I. Yotov. A parallel multiblock/multidomain approach to reservoir simulation. In *Fifteenth SPE Symposium on Reservoir Simulation, Houston, Texas*, pages 51–62. Society of Petroleum Engineers, 1999. SPE 51884.

D. Wildenschild, J. Hopmans, C. Vaz, M. Rivers, and D. Rikard. Using X-ray computed tomography in hydrology. Systems, resolutions, and limitations. *Journal of Hydrology*, 267:285–297, 2002.

I. Yotov. Advanced techniques and algorithms for reservoir simulation IV. Multiblock solvers and preconditioners. In J. Chadam, A. Cunningham, R. E. Ewing, and M. F. Ortoleva, P. Wheeler, editors, *IMA Volumes in Mathematics and its Applications, Volume 131: Resource Recovery, Confinement, and Remediation of Environmental Hazards*. Springer, 2002.

L. Young. A study for simulating fluid displacements in petroleum reservoirs. *Computational Methods and Application in Mechanical Engineering*, 47:3–46, 1984.

H. Zhang and F.W. Schwartz. Simulating the *in situ* oxidative treatment of chlorinated compounds by potassium permanganate. *Water Resources Research*, 36(10): 3031–3042, 2000.

Die VDM Verlagsservicegesellschaft sucht für wissenschaftliche Verlage abgeschlossene und herausragende

Dissertationen, Habilitationen, Diplomarbeiten, Master Theses, Magisterarbeiten usw.

für die kostenlose Publikation als Fachbuch.

Sie verfügen über eine Arbeit, die hohen inhaltlichen und formalen Ansprüchen genügt, und haben Interesse an einer honorarvergüteten Publikation?

Dann senden Sie bitte erste Informationen über sich und Ihre Arbeit per Email an *info@vdm-vsg.de*.

Sie erhalten kurzfristig unser Feedback!

VDM Verlagsservicegesellschaft mbH
Dudweiler Landstr. 99   Telefon +49 681 3720 174
D - 66123 Saarbrücken   Fax   +49 681 3720 1749
www.vdm-vsg.de

Die VDM Verlagsservicegesellschaft mbH vertritt

Printed by Books on Demand GmbH, Norderstedt / Germany